CONSTRUCTION ESTIMATING FOR BEGINNERS

fundamentals for accurate and efficient estimates

Steven Smith, Ph.D.

Wisdom Publishers

Copyright © 2024 Wisdom Publishers

All rights reserved

No part of this book may be reproduced, or stored in a retrieval system, or transmitted in any form or by any means, electronic, mechanical, photocopying, recording, or otherwise, without express written permission of the publisher.

ISBN: 9798326800428
Imprint: Independently published

Cover design by: Art Painter
Library of Congress Control Number: 2018675309
Printed in the United States of America

To the dreamers and tireless seekers of knowledge, the builders of tomorrow who envision and create, may this book serve as your guiding light on your journey to mastering the craft of construction estimating.

To my father, a skilled mason whose calloused hands and boundless knowledge first sparked my passion for construction. His legacy of dedication and craftsmanship continues to inspire me. And to my wife, Sarah, your unwavering support has been the fuel that propelled this endeavor forward. You are my constant source of strength and encouragement. This book is dedicated to you both, with all my heart.

CONTENTS

Title Page
Copyright
Dedication
Foreword
Introduction
Preface

Part 1: Essential foundation of construction estimating	1
Chapter 1: Introduction to construction estimating	2
Chapter 2: Construction documents and specifications	6
Chapter 3: Mastering takeoffs	17
Part 2: Building your estimate	32
Chapter 4: Cost estimating methods for beginners	33
Chapter 5: Labor estimating essentials	44
Chapter 6: Material estimating fundamentals	53
Part 3: Refining estimate for accuracy and efficiency	60
Chapter 7: Overhead and profit considerations	61
Chapter 8: Risk management and contingency planning	71
Chapter 9: Presenting and communicating estimate effectively	80
Part 4: Mastering efficiency and technology in the modern era	89

Chapter 10: Understanding construction estimating software	90
Part 5: Putting it all together: practical applications	97
Chapter 11: Bidding and project management integration	98
Part 6: Beyond the basics	106
Chapter 12: Advanced estimating techniques	107
Chapter 13: Staying ahead of the curve	115
Appendix	124
Acknowledgement	133
About The Author	135

FOREWORD

In the fast-paced world of construction, getting your estimates right is crucial for project success. After more than twenty years in this industry, I've witnessed firsthand how accurate estimating can make or break a project. This book is a fantastic guide for anyone new to construction estimating. I consider it a valuable resource for students, those just starting out, and even experienced professionals who want to sharpen their skills.

The author has done an excellent job of breaking down complicated ideas into simple, practical advice. You'll find clear explanations, practical examples, and useful information that you can start using right away. Estimating in construction isn't just about adding up costs. It's about understanding the details of a project, predicting possible issues, and making smart decisions. This book will guide you through the entire process, helping you develop the skills you need to create accurate and efficient estimates.

Learning to estimate well is both a science and an art. It takes practice, a willingness to learn, and attention to detail. This book encourages you to take on these challenges, learn from your experiences, and aim for accuracy and efficiency in every estimate you make.

As a practicing Quantity Surveyor with over 20 years of

experience, I understand the challenges faced by construction professionals at all levels. This book is the guide I wish I had when I started, and I believe it will be a valuable asset to yours as well. I believe it will be a valuable tool for your professional growth and success in construction estimating. It's a reliable resource that will support you as you build your career in this important field.

As an expert in the industry, I can confidently say that this book is a must-read. It provides the essential knowledge and practical advice you need to handle construction estimating with confidence and skill. May you succeed as you make the most of this book.

Sincerely,

Nathan Anderson

INTRODUCTION

Join Steven Smith as he navigates the world of construction estimating in 'Construction Estimating for Beginners: fundamentals for accurate and efficient estimates.' This book offers clear and practical insights in the field of construction estimating, providing essential knowledge and techniques to excel in the art and science of estimating.

PREFACE

Welcome to "Construction Estimating for Beginners: fundamentals for accurate and efficient estimates." This book is primarily written for individuals who are new to the field of construction estimating, providing a solid foundation in the essential principles and practices required to produce accurate and efficient estimates. It can also serve as a valuable resource for construction professionals, including contractors, estimators, project managers, and students, who are involved in the process of estimating costs for construction projects.

Overview of the book's purpose and audience

The purpose of this book is to demystify construction estimating and empower beginners with the knowledge and skills needed to succeed in this critical aspect of the construction industry. It is designed to meet the needs of students exploring career opportunities, aspiring estimators looking to enter the workforce, and construction professionals transitioning into estimating.

How to best use this book for practical learning and application

This book is structured to guide beginners through the

fundamentals of construction estimating in a practical and accessible manner. In the following ways, you can make the most of your learning experience:

- **Start with the basics**: Begin by familiarizing yourself with the foundational concepts of construction estimating presented in the early chapters of the book. Take your time to grasp these fundamental principles before moving on to more advanced topics.

- **Engage actively with the material**: Actively engage with the content by taking notes, asking questions, and seeking clarification on any concepts that may be unclear. This interactive approach will enhance your understanding and retention of the material.

- **Apply what you learn**: Apply the concepts and techniques learned in each chapter to practical exercises and real-world scenarios. Practice performing takeoffs, estimating costs, and developing bid proposals to reinforce your learning and build confidence in your abilities.

- **Utilize resources and tools**: Take advantage of the resources and tools provided in the book, including templates and examples. These resources are designed to assist you in applying theoretical concepts to practical situations and honing your estimating skills.

- **Seek continuous improvement**: Recognize that construction estimating is a skill that develops over time with practice and experience. Stay curious, remain open to learning opportunities, and actively seek feedback to continuously improve your estimating abilities.

Adhering to these guidelines and approaching your learning journey with enthusiasm and dedication, you will lay a solid foundation for success in construction estimating. Remember, every step you take brings you closer to mastering the fundamentals and becoming a proficient estimator.

PART 1: ESSENTIAL FOUNDATION OF CONSTRUCTION ESTIMATING

CHAPTER 1: INTRODUCTION TO CONSTRUCTION ESTIMATING

Construction estimating is the process of calculating the costs associated with a construction project. It involves assessing various factors such as materials, labor, equipment, and overhead expenses to determine the overall cost of completing the project. This estimate serves as a crucial guideline for budgeting, bidding, and planning throughout the project's lifecycle.

The importance of construction estimating cannot be overstated for several reasons:

1. **Budgeting and financial planning**: Accurate estimating helps stakeholders develop realistic budgets for construction projects. By forecasting costs upfront, they can allocate resources efficiently, minimize financial risks, and ensure projects remain financially viable.

2. **Competitive bidding**: Estimating plays a pivotal role

in the bidding process. Contractors rely on accurate estimates to prepare competitive bids that reflect the true cost of the project. A well-calculated estimate increases the contractor's chances of winning the bid while maintaining profitability.
3. **Project feasibility**: Before embarking on a construction project, stakeholders need to assess its feasibility from a financial perspective. Estimating allows them to evaluate whether the project aligns with their budgetary constraints and investment objectives.
4. **Resource allocation**: Estimating helps allocate resources effectively by determining the quantity and type of materials, equipment, and labor required for the project. This ensures that resources are utilized optimally, minimizing waste and maximizing productivity.
5. **Risk management**: Accurate estimating enables stakeholders to identify and mitigate potential risks associated with the project. By anticipating and budgeting for unforeseen expenses or delays, they can proactively manage risks and maintain project timelines and budgets.
6. **Client satisfaction**: Reliable estimates instill confidence in clients and stakeholders by providing transparency and clarity regarding project costs. Clients are more likely to trust contractors who can deliver accurate estimates, leading to stronger relationships and increased client satisfaction.

Types of construction estimates and their uses

1. **Preliminary estimates**

- **Overview**: Preliminary estimates are rough approximations of project costs used during the early stages of project development.
- **Uses**: They help stakeholders assess project feasibility, establish initial budgets, and make high-level decisions regarding project scope and viability.

2. Detailed estimates

- **Overview**: Detailed estimates provide comprehensive breakdowns of project costs based on a thorough analysis of project requirements and specifications.
- **Uses**: They are utilized during the design phase to develop accurate budgets, prepare competitive bids, and inform decision-making regarding project planning and execution.

3. Unit cost estimates

Overview: Unit cost estimates involve estimating costs based on predetermined unit prices for materials, labor, and equipment.

Uses: They are commonly used for repetitive tasks or standard components of construction projects, enabling quick and efficient cost calculations.

4. Assembly estimates

- **Overview**: Assembly estimates focus on estimating costs for pre-defined assemblies or systems within a construction project, such as HVAC systems or structural components.
- **Uses**: They streamline the estimating process by grouping similar elements, allowing for more accurate and efficient cost calculations for complex project components.

5. **Parametric estimates:**

- **Overview**: Parametric estimates utilize historical data and project parameters to predict costs based on key project metrics, such as square footage or building height.

- **Uses**: They provide quick and preliminary cost estimates based on project characteristics, allowing stakeholders to assess project feasibility and make initial budgetary decisions.

Each type of construction estimate serves a specific purpose and is utilized at different stages of the project lifecycle. By understanding the characteristics and applications of each type, stakeholders can develop more accurate and reliable estimates to support informed decision-making and successful project execution.

CHAPTER 2: CONSTRUCTION DOCUMENTS AND SPECIFICATIONS

Understanding construction documents is essential for accurate estimating. These documents include plans, specifications, and contracts that detail every aspect of a project, from design to execution.

Understanding construction documents

1. Plans

- **Overview**: Plans, also known as blueprints or drawings, are graphical representations of the project design created by architects, engineers, and designers.

- **Types**: Plans can include architectural drawings, structural drawings, mechanical drawings, electrical drawings, plumbing drawings, and landscape drawings.

- **Contents**: They depict the layout, dimensions, and details of various elements of the project, such as floor plans, elevations, sections, details, and schedules.
- **Relevance to estimating**: Estimators analyze plans to determine quantities of materials, labor requirements, and costs associated with various project components.
- **Quantification**: Estimators perform takeoffs or quantity surveys based on plans to quantify materials needed for construction, such as concrete, steel, lumber, piping, wiring, and finishes.
- **Cost estimation**: By associating unit costs with the quantities derived from plans, estimators calculate the total cost of materials and labor required to execute the project according to the specified design.

2. Specifications

- **Overview**: Specifications, often referred to as specs, are written documents that detail the materials, products, standards, and workmanship required for the construction project.
- **Contents**: Specifications specify the quality, performance, and installation requirements for various construction elements, including materials, finishes, equipment, and systems. They may also include testing, inspection, and certification requirements.
- **Formats**: Specifications are typically organized by divisions or sections, following standardized formats such as MasterFormat or Uniformat.
- **Relevance to estimating**: Estimators review specifications to understand the quality and performance criteria that influence material selection and pricing.

- **Material selection**: Estimators consider specifications when selecting materials and products for the project, ensuring compliance with quality standards, building codes, and client preferences.
- **Cost analysis**: Estimators assess the cost implications of different material options specified in the project, comparing prices, availability, and performance characteristics to determine the most cost-effective solutions.

3. Contracts

- **Overview**: Contracts are legal agreements between project stakeholders, outlining the rights, obligations, responsibilities, and terms and conditions governing the construction project.
- **Parties**: Contracts may involve multiple parties, including the owner or client, general contractor, subcontractors, suppliers, architects, engineers, consultants, and other stakeholders.
- **Contents**: Contracts specify the scope of work, project schedule, payment terms, insurance requirements, dispute resolution mechanisms, change order procedures, warranties, and other contractual provisions.
- **Relevance to estimating**: Contracts define the scope of work, terms, and conditions of the construction project, establishing the basis for estimating costs and preparing bids.
- **Scope definition**: Estimators refer to contracts to understand project requirements, exclusions, and deliverables, ensuring that estimates accurately reflect

the agreed-upon scope of work.

- **Bid preparation**: Estimators use contract documents to develop bid proposals that align with the project's contractual obligations, pricing strategies, payment terms, and risk management considerations.

- **Legal compliance**: Estimators ensure that estimates adhere to contractual requirements, such as pricing transparency, accuracy, and consistency, to mitigate the risk of disputes or claims during project execution.

Understanding and interpreting these documents accurately is fundamental to the estimating process. Effective estimators must be adept at reading plans, understanding specifications, and navigating contracts to create precise and comprehensive estimates that support successful project execution.

How to read and interpret construction documents for estimating

Reading and interpreting construction documents is a crucial skill for accurate estimating. These documents contain the vital details needed to develop a comprehensive and reliable estimate. This section provides you with a step-by-step guide to effectively read and interpret construction documents:

1. Familiarize with the types of documents

- **Plans/blueprints**: Include architectural, structural, mechanical, electrical, and plumbing drawings.

- **Specifications**: Detailed descriptions of materials, workmanship, and standards.

- **Contracts**: Legal agreements outlining scope, responsibilities, terms, and conditions.

2. Start with the overall view

- **Project overview**: Begin by understanding the general scope and objectives of the project. This will provide context for more detailed examinations.

- **Cover sheet**: Review the cover sheet of the plans, which often includes project title, location, key contacts, and an index of all drawings.

3. Understand the organization of documents

- **Plan set organization**: Plans are typically organized by discipline (e.g., architectural, structural) and by level of detail (e.g., overall site plans, detailed sections).

- **Specification divisions**: Specifications are usually divided into sections based on the CSI MasterFormat or similar standards, covering various aspects of construction.

4. Interpret the plans

- **Architectural plans**: Examine floor plans, elevations, and sections to understand the building layout, dimensions, and key architectural features.

- **Floor plans**: Identify room sizes, wall placements, door and window locations.

- **Elevations**: View the exterior faces of the building, noting materials and finishes.

- **Sections**: Look at cut-through views of the building to understand heights and structural elements.

- **Structural plans**: Review details on foundations, framing, beams, and columns.

- **MEP plans**: Study mechanical, electrical, and plumbing drawings for system layouts, equipment specifications, and routing.

5. Extract key information

- **Quantities**: Identify and measure dimensions to calculate quantities of materials needed. Use tools like scaling, manual takeoffs, or digital takeoff software.

- **Materials**: Note the types and specifications of materials called out in the drawings and specs.

- **Labor**: Determine labor requirements by understanding the complexity and scale of the tasks shown in the plans.

6. Cross-reference specifications

- **Material specifications**: Cross-check the plans with the specifications to verify material types, grades, and installation methods.

- **Quality standards**: Ensure that the quality standards outlined in the specifications are aligned with the details in the plans.

- **Execution requirements**: Review specifications for execution standards, tolerances, and testing requirements.

7. Navigate the contracts

- **Scope of work**: Clarify the full extent of the work to be performed as outlined in the contract documents.

- **Terms and conditions**: Understand payment terms, timelines, change order procedures, and other critical contract terms.

- **Responsibilities**: Identify the responsibilities of each party involved in the project.

8. Use checklists and annotations

- **Checklists**: Develop checklists for reviewing each type of document to ensure no details are overlooked.
- **Annotations**: Make notes directly on copies of the plans or use software tools to annotate digital versions for easy reference.

9. Leverage technology

- **Digital takeoff tools**: Use digital tools to perform accurate and efficient quantity takeoffs.
- **BIM software**: Utilize Building Information Modeling to visualize project components in 3D and extract detailed information.

10. Collaborate with stakeholders

- **Clarifications**: Communicate with architects, engineers, and other stakeholders to clarify any ambiguities or discrepancies in the documents.
- **Updates**: Stay informed about any changes or updates to the plans, specifications, or contracts that could impact the estimate.

By mastering the ability to read and interpret construction documents, estimators can ensure they capture all necessary details, produce accurate cost estimates, and contribute to the successful planning and execution of construction projects.

Identifying key information for estimating

Accurate estimating hinges on the ability to identify and extract key information from construction documents. This section explains how to identify the crucial details for quantities, materials, and labor.

1. Quantities

a. Reviewing plans

- **Floor plans**: Measure the dimensions of rooms, walls, and other elements to calculate the areas and volumes. Pay attention to scale and use tools like scaling rulers or digital takeoff software.
- **Elevations and sections**: Calculate the heights and cross-sectional areas for various elements to determine quantities of vertical components like walls, columns, and facades.
- **Schedules**: Use window, door, and finish schedules to count the number of specific items required.

b. Performing takeoffs

- **Manual takeoffs**: Use scaling tools to manually measure dimensions on printed plans. Record measurements systematically.
- **Digital takeoffs**: Employ software to digitize the plans and automate the measurement process. This improves accuracy and efficiency.

c. Quantification examples

- **Concrete**: Calculate volumes by measuring the length, width, and depth of footings, slabs, and columns.
- **Steel**: Measure the lengths and weights of beams, rebar, and structural steel members.
- **Drywall**: Determine the square footage of wall and ceiling areas to be covered.

2. Materials

a. Reviewing specifications

- **Material types**: Identify the types of materials specified for different components of the project, including their grades, finishes, and performance criteria.
- **Quality standards**: Note any specified standards or certifications that materials must meet, such as ASTM, ANSI, or local building codes.

b. Cross-referencing plans and specifications

- **Consistency check**: Ensure that the materials detailed in the specifications match those shown in the plans. For example, if a specific type of insulation is specified, confirm its location and quantity on the plans.
- **Material lists**: Create comprehensive lists of materials needed, including quantities and specifications.

c. Material examples

- **Concrete mixes**: Identify the required mix design, strength, and any additives or reinforcements.
- **Finishes**: Note the types of finishes for walls, floors, and ceilings, including paint, tile, carpet, and other surface treatments.
- **Mechanical and electrical components**: Identify specifications for HVAC equipment, lighting fixtures, wiring, and plumbing fixtures.

3. Labor

a. Understanding labor requirements

- **Labor tasks**: Break down the project into specific tasks, such as framing, electrical work, plumbing, and finishing. Each task requires a different set of skills and labor categories.

- **Productivity rates**: Use historical data or industry standards to determine the productivity rates for various tasks. This includes the amount of work a laborer or crew can complete in a given period.

b. **Reviewing plans and specifications for labor**

- **Complexity and detail**: Assess the complexity of construction details to estimate labor effort. More complex details typically require more skilled labor and longer timeframes.

- **Special requirements**: Identify any special labor requirements, such as certified tradespeople for electrical work or specialized installers for certain finishes.

c. **Labor estimating examples**

- **Framing**: Calculate the number of carpenters needed based on the amount of framing required and the productivity rate for framing tasks.

- **Electrical work**: Estimate the number of electricians required by calculating the length of conduit runs, the number of outlets, and the complexity of the electrical system.

- **Plumbing**: Determine the number of plumbers needed by assessing the number of fixtures, the complexity of the plumbing network, and the productivity rates for installation.

4. Integrating quantities, materials, and labor

a. Cost estimation

- **Unit costs**: Apply unit costs to the quantities of materials and labor determined from the plans and specifications. Unit costs should include both direct costs (materials and labor) and indirect costs (overhead, equipment, etc.).
- **Subcontractor quotes**: Obtain quotes from subcontractors for specialized work to ensure accuracy in labor and material costs.

b. Finalizing the estimate

- **Detailed estimates**: Compile all quantified data, material specifications, and labor requirements into a detailed estimate. Ensure all costs are accounted for and that the estimate aligns with the project scope and specifications.

- **Review and adjust**: Review the estimate for accuracy, consistency, and completeness. Adjust as necessary based on feedback from stakeholders or new information.

Identifying and analyzing the key information related to quantities, materials, and labor, helps estimators develop reliable construction estimates. This thorough approach helps ensure that projects are planned accurately, budgeted properly, and executed successfully.

CHAPTER 3: MASTERING TAKEOFFS

A takeoff in construction estimating refers to the process of quantifying materials and labor required for a project based on a detailed examination of construction documents such as plans and specifications. Takeoffs play a pivotal role in estimating as they form the foundation for determining the quantities and costs associated with various project components. Here's a comprehensive overview of takeoffs and their significance in the estimating process:

1. Definition of takeoff

A takeoff involves systematically reviewing construction documents to identify and measure the quantities of materials, equipment, and labor needed to complete a project. It entails extracting specific details from plans, specifications, and other relevant documents to create a comprehensive list of items required for construction.

2. Role of takeoffs in estimating

a. Quantifying materials

Takeoffs enable estimators to accurately quantify the amounts of materials needed for different aspects of the project, such as structural elements, finishes, and mechanical systems.

By measuring dimensions and calculating areas, volumes, or lengths from the plans, estimators can determine the quantities of concrete, steel, lumber, drywall, piping, wiring, and other materials required.

b. Estimating labor requirements

In addition to materials, takeoffs also help estimators assess the labor requirements for various construction tasks.

By identifying components and assemblies on the plans, estimators can determine the labor hours and skill levels needed to complete each task, considering factors such as complexity, size, and sequencing.

c. Determining equipment needs

Takeoffs may also involve identifying the equipment and machinery required for construction activities, such as cranes, excavators, or specialized tools.

Estimators can assess the types and quantities of equipment needed based on the project scope, site conditions, and construction methods outlined in the plans.

d. Cost estimation

Once the quantities of materials, labor, and equipment are determined through the takeoff process, estimators assign unit costs to each item to calculate the total cost of the project.

The takeoff serves as the basis for generating a detailed cost estimate, which includes direct costs (materials, labor, equipment) and indirect costs (overhead, profit, contingency).

e. Supporting decision-making

Takeoffs provide valuable insights into the scope and scale of the project, helping stakeholders make informed decisions regarding budgeting, scheduling, and resource allocation.
Accurate takeoffs ensure that estimates reflect the true requirements of the project, reducing the risk of cost overruns and delays during construction.

3. Methods of performing takeoffs

a. Manual takeoffs

Estimators manually measure dimensions and quantities from printed or digital construction documents using scaling tools, rulers, or digitized overlays. They may use worksheets or spreadsheets to record measurements and perform calculations for each item.

b. Digital takeoff software

Estimators utilize specialized software applications designed for construction takeoffs, which streamline the process and improve accuracy and efficiency. Digital takeoff tools allow for electronic measurement directly on digital plans, automatic calculation of quantities, and integration with estimating software for seamless cost analysis.

4. Importance of accuracy and detail

Accurate takeoffs are essential for producing reliable cost estimates and ensuring that projects are adequately funded and resourced. Estimators must pay close attention to detail, carefully reviewing plans and specifications to avoid errors or omissions that could impact the accuracy of the estimate.

Takeoffs are a fundamental aspect of construction estimating, providing the detailed quantities of materials, labor, and equipment needed to develop comprehensive cost estimates for construction projects. By accurately quantifying project requirements through the takeoff process, estimators play a vital role in facilitating informed decision-making and successful project outcomes.

Different types of takeoffs (manual, software-based)

Takeoffs in construction estimating can be performed using various methods, each with its own advantages and considerations. In this section, we'll explore two primary types of takeoffs: manual takeoffs and software-based takeoffs, highlighting their characteristics, benefits, and limitations.

1. Manual takeoffs

Characteristics

- **Traditional approach**: Manual takeoffs involve physically measuring dimensions and quantities from printed construction documents such as plans and specifications.

- **Analog tools**: Estimators use tools like rulers, scaling devices, digitized overlays, and printed worksheets to record measurements and perform calculations.

- **Labor-intensive**: Manual takeoffs require significant time and effort, as each item must be measured individually, and calculations are performed manually.

Benefits

- **Low cost**: Manual takeoffs typically require minimal investment in equipment or software, making them accessible to small firms or individual estimators with limited resources.
- **Hands-on approach**: Estimators have direct control over the measurement process and can visually inspect construction documents for accuracy and detail.
- **Flexibility**: Manual takeoffs can be performed in any location without reliance on digital technology or internet connectivity.

Limitations

- **Time-consuming**: Manual takeoffs are labor-intensive and time-consuming, especially for large or complex projects, leading to slower turnaround times for estimates.
- **Prone to errors**: Human error is inherent in manual processes, increasing the risk of measurement inaccuracies, calculation mistakes, or omissions.
- **Limited scalability**: Manual takeoffs may not be practical for projects with tight deadlines or high volume of work due to their slower pace and limited scalability.

2. Software-based takeoffs

Characteristics

- **Digital approach**: Software-based takeoffs leverage specialized applications or tools designed for construction estimating, facilitating electronic measurement and calculation processes.
- **Digitized plans**: Estimators work with digital versions of construction documents, such as PDFs or CAD files, which can be viewed, measured, and annotated electronically.
- **Automation**: Software tools automate measurement tasks, allowing for faster quantification of quantities and seamless integration with estimating software.

Benefits

- **Efficiency**: Software-based takeoffs offer significant time savings compared to manual methods, as measurements can be performed quickly and accurately with digital tools.
- **Accuracy**: Automation reduces the risk of human error, ensuring precise measurements and calculations for each item in the takeoff.
- **Scalability**: Digital tools are well-suited for large or complex projects, as they can handle high volumes of data and streamline repetitive tasks.

Limitations

- **Learning curve**: Users may require training to effectively utilize software-based takeoff tools, especially if they are unfamiliar with digital construction documentation and measurement techniques.
- **Cost**: Software licenses and subscriptions for specialized takeoff software may incur upfront costs and ongoing expenses, which could be prohibitive for smaller firms or

individual estimators.

- **Dependency on technology**: Software-based takeoffs rely on access to digital devices, software applications, and reliable internet connectivity, which may not always be available in remote or field environments.

Choosing the right approach

- **Project size and complexity**: Consider the scale and complexity of the project when selecting a takeoff method. Manual takeoffs may be sufficient for smaller projects with straightforward requirements, while software-based takeoffs are better suited for larger, more intricate projects.

- **Budget and resources**: Evaluate the available budget and resources, including personnel, equipment, and software, to determine the most cost-effective approach for conducting takeoffs.

- **Efficiency vs. accuracy**: Balance the need for efficiency with the requirement for accuracy. While software-based takeoffs offer speed and automation, manual takeoffs provide greater control and attention to detail.

Both manual and software-based takeoffs have their place in construction estimating, offering distinct advantages and considerations. Estimators should carefully evaluate project requirements, budget constraints, and personnel capabilities to determine the most appropriate method for conducting takeoffs and producing accurate cost estimates.

Performing accurate quantity takeoffs from construction documents

Performing accurate quantity takeoffs from construction documents is essential for developing reliable cost estimates and effectively managing construction projects. Here's a detailed guide on how to conduct accurate quantity takeoffs:

1. Gather construction documents

a. Plans

- Obtain architectural, structural, mechanical, electrical, and plumbing drawings for the project.
- Ensure that the plans are complete, legible, and accurately represent the scope of work.

b. Specifications

- Review the project specifications to understand material requirements, quality standards, and installation methods.

c. Other documents

- Gather any additional documents relevant to the project, such as schedules, scopes of work, or addenda.

2. Understand the scope of work

a. Project overview

- Familiarize yourself with the overall scope and objectives of the project.

- Identify key components and areas that require quantity takeoffs.

b. Breakdown tasks

- Divide the project into manageable tasks or work packages, such as site work, foundations, structural framing, finishes, and MEP systems.

3. Establish measurement standards

a. Units of measurement

- Determine the appropriate units of measurement for each type of material or component (e.g., square feet for flooring, linear feet for piping).

b. Consistency

- Ensure consistency in measurement units across all takeoffs to avoid errors or confusion.

4. Perform quantity takeoffs

a. Manual takeoffs

- Use rulers, scaling devices, or digitized overlays to measure dimensions directly on printed plans.
- Record measurements systematically, noting the quantities of materials for each component.

b. Digital takeoff software

- Utilize specialized software tools for digital takeoffs, which allow for electronic measurement and automated quantity calculations.
- Import digital plans into the software and use tools like area, length, and count functions to measure quantities accurately.

5. Break down components

a. Components and assemblies

- Identify and quantify individual components and assemblies depicted on the plans, such as walls, doors, windows, structural members, fixtures, and equipment.

b. Segregate areas

- Break down areas into smaller sections for more accurate measurement, especially for irregularly shaped spaces or complex layouts.

6. Account for wastage and allowances

a. Waste factors

- Consider factors for material wastage, cutting losses, and spoilage when calculating quantities. Use industry-standard waste factors or project-specific allowances based on experience and project conditions.

b. Contingency allowances

- Include contingency allowances in the quantities to account for unforeseen changes or variations during construction.

7. Cross-reference with specifications

a. Material specifications

- Cross-reference the quantities derived from the plans with the material specifications to ensure compliance with quality standards and performance criteria.

b. Quality checks

- Verify that the specified materials and installation methods align with the project requirements and regulatory standards.

8. Review and validate

a. Check for errors

- Review the quantity takeoffs for accuracy, ensuring that all measurements are correct and consistent.
- Double-check calculations and conversions to minimize errors.

b. Peer review

- Have a colleague or supervisor review the quantity takeoffs to validate the accuracy and completeness of the estimates.

9. Document and organize

a. Record keeping

- Document the quantity takeoffs systematically, maintaining detailed records of measurements, calculations, and assumptions.

- Organize the information in a structured format for easy reference and retrieval.

b. Version control

- Maintain version control of the quantity takeoffs to track any revisions or updates throughout the estimating process.

10. Communicate and collaborate

a. Stakeholder engagement

- Communicate the quantity takeoff findings effectively with project stakeholders, including clients, designers, contractors, and suppliers.
- Address any questions or concerns and seek clarification on discrepancies or ambiguities in the construction documents.

b. Collaboration tools

- Utilize collaboration tools or platforms to facilitate communication and coordination among project team members involved in the estimating process.

By following these steps and best practices, you can perform accurate quantity takeoffs from construction documents, providing a solid foundation for developing reliable cost estimates and effectively managing construction projects from inception to completion.

Common tools and software for takeoffs

Several tools and software options are available for performing takeoffs in construction estimating, catering to different preferences, project requirements, and budget considerations. The following are some common tools and software used for takeoffs:

1. Manual takeoff tools

a. Measuring tools

- **Architectural scale ruler**: Used for measuring dimensions directly on printed plans.
- **Digital calipers**: Precise measurement tool for accurately determining lengths, widths, and thicknesses.
- **Tape measure**: Handy for measuring larger dimensions or distances on physical surfaces.

b. Digitized overlays

- **Tracing paper**: Allows for tracing components or areas of interest on printed plans for further analysis or calculation.
- **Transparent grid sheets**: Overlay sheets with grids or scales for measuring and scaling dimensions on plans.

2. Digital takeoff software

a. Standalone takeoff software

- **Bluebeam revu**: Comprehensive software for PDF-based takeoffs, offering measurement tools, markup capabilities, and integration with estimating software.
- **PlanSwift**: Specialized takeoff software with features for electronic measurement, area calculations, and customizable templates.

- **On-Screen takeoff**: Digital takeoff solution for measuring quantities directly on digital plans, with integration options for transferring data to estimating software.

b. Estimating software with built-in takeoff features

- **ProEst**: Estimating software with integrated takeoff functionality for generating detailed cost estimates and bid proposals.

- **Sage estimating**: Construction estimating software with advanced takeoff capabilities, including assembly-based estimating and database integration.

- **STACK estimating**: Cloud-based estimating platform with takeoff tools for measuring quantities, creating material lists, and collaborating on projects online.

3. Building information modeling software

- **Autodesk revit**: BIM software for modeling, designing, and documenting building projects, with capabilities for quantifying materials and generating schedules directly from the BIM model.

- **Trimble tekla structures**: Structural BIM software with tools for modeling steel, concrete, and precast structures, including features for estimating quantities and generating accurate bills of materials.

- **Graphisoft archicad**: BIM solution for architects and designers, offering integrated quantity takeoff tools for analyzing building components and generating material lists.

4. Spreadsheets and database software

- **Microsoft excel**: Spreadsheet software commonly used for organizing, calculating, and managing quantity takeoff data, with customizable templates for various construction trades.

- **Access database**: Database software for creating centralized repositories of project information, including quantities, materials, labor rates, and cost data.

Considerations for choosing the right tool

Project requirements: Select a tool or software that aligns with the scope, complexity, and scale of the project.

User experience: Consider the ease of use, learning curve, and familiarity of the tool or software for the estimating team.

Integration: Evaluate integration capabilities with other software systems, such as estimating, project management, or accounting software.

Cost: Determine the upfront costs, licensing fees, and ongoing expenses associated with the tool or software, considering the budget constraints of the organization.

By leveraging these common tools and software options for takeoffs, construction estimators can streamline the quantity measurement process, improve accuracy, and efficiently generate comprehensive cost estimates for construction projects.

PART 2: BUILDING YOUR ESTIMATE

CHAPTER 4: COST ESTIMATING METHODS FOR BEGINNERS

Unit cost approach

The unit cost approach is a straightforward method used to estimate the cost of construction projects by applying pre-established unit prices to specific quantities of materials, labor, or equipment. This method is particularly useful for beginners in construction estimating as it provides a simplified framework for developing cost estimates without extensive experience or detailed analysis.

1. Understanding unit costs

- **Definition**: Unit costs refer to the price per unit of measurement for various construction elements, such as materials, labor, or equipment.
- **Examples**: Unit costs can be expressed as cost per square foot, cost per linear foot, cost per cubic yard, or cost per hour, depending on the item being estimated.
- **Sources**: Unit costs can be obtained from industry-

standard reference books, cost databases, historical project data, or supplier quotations.

2. Application of unit costs

- **Material costs**: Estimators multiply the quantity of materials required for the project by the unit cost per unit of measurement (e.g., cost per square foot for flooring or cost per cubic yard for concrete).
- **Labor costs**: Labor costs are calculated by multiplying the estimated labor hours for specific tasks by the unit cost per hour for labor, considering factors such as skill level and wage rates.
- **Equipment costs**: Estimators determine equipment costs by multiplying the estimated hours of equipment usage by the unit cost per hour for each type of equipment required for the project.

3. Advantages of the unit cost approach

- **Simplicity**: The unit cost approach offers a straightforward and easy-to-understand method for beginners to develop cost estimates without complex calculations.
- **Time savings**: By relying on pre-determined unit prices, estimators can quickly generate cost estimates without the need for extensive analysis or detailed measurements.
- **Consistency**: Standardized unit prices promote consistency across different projects, ensuring uniformity in estimating practices and results.

4. Limitations of the unit cost approach

- **Accuracy**: Unit costs may not always reflect the specific conditions or requirements of a particular project, leading to inaccuracies in the estimate.

- **Scope limitations**: The unit cost approach may be less suitable for projects with unique or specialized elements that cannot be adequately represented by standardized unit prices.
- **Dependency on data quality**: The accuracy of cost estimates derived from the unit cost approach depends on the quality and reliability of the unit cost data used.

5. **Best practices for using unit costs**

- **Validation**: Validate unit costs by comparing them to current market rates, supplier quotations, or historical project data to ensure relevance and accuracy.
- **Adjustment**: Adjust unit costs as needed to account for project-specific factors, such as location, project size, complexity, or material specifications.
- **Documentation**: Maintain a record of the unit costs used in the estimate, including their sources and any adjustments made, to support transparency and accountability.

Beginners in construction estimating can use the unit cost approach to develop initial cost estimates efficiently and effectively, laying the groundwork for further learning and refinement of estimating techniques.

Assembly cost estimating

Assembly cost estimating is a method where costs are estimated based on pre-defined assemblies or components rather than individual materials or tasks. This approach simplifies the estimating process by grouping related items together, offering beginners a systematic framework for developing cost estimates.

1. **Understanding assemblies**

    - **Definition**: Assemblies are predefined groupings of materials, components, and labor activities that represent common construction elements or building systems.

    - **Examples**: Assemblies can include items such as wall assemblies, roof assemblies, window assemblies, door assemblies, or HVAC systems.

    - **Characteristics**: Each assembly typically consists of a list of materials, quantities, labor activities, and associated costs required to complete the assembly.

2. **Application of assembly cost estimating**

    - **Selection of assemblies**: Begin by identifying and selecting relevant assemblies based on the project scope, specifications, and typical construction practices.

    - **Quantification**: Determine the quantities of materials, labor hours, and equipment required for each assembly based on project requirements and specifications.

    - **Cost Calculation**: Calculate the total cost of each assembly by multiplying the quantities of materials and labor by their respective unit costs or rates.

    - **Aggregation**: Summarize the costs of individual assemblies to derive the total project cost estimate.

3. **Advantages of assembly cost estimating**

    - **Efficiency**: Assembly cost estimating streamlines the estimating process by grouping related items together, reducing the time and effort required to develop cost estimates.

    - **Accuracy**: By using pre-defined assemblies with

standardized components and labor activities, assembly cost estimating promotes consistency and accuracy in estimating project costs.

- **Simplicity**: This method is particularly suitable for beginners as it offers a structured and systematic approach to estimating costs without the need for extensive analysis or detailed measurements.

4. Limitations of assembly cost estimating

- **Scope limitations**: Assembly cost estimating may not be suitable for projects with unique or custom elements that cannot be adequately represented by pre-defined assemblies.

- **Flexibility**: The use of pre-defined assemblies may limit the flexibility to accommodate changes or variations in project requirements, potentially leading to inaccuracies in the estimate.

- **Dependency on assemblies**: The accuracy of the estimate relies on the availability of appropriate assemblies and their alignment with the project scope and specifications.

5. Best practices for using assembly cost estimating

- **Selection**: Choose assemblies that closely match the project scope, specifications, and construction practices to ensure relevance and accuracy.

- **Validation**: Validate assembly costs by comparing them to industry standards, historical data, or supplier quotations to verify their accuracy and appropriateness.

- **Documentation**: Maintain documentation of selected assemblies, including their components, quantities, and costs, to support transparency and facilitate future reference or adjustments.

Parametric estimating

Parametric estimating is a technique that uses historical data and project characteristics to estimate costs based on predetermined parameters or formulas. This method involves identifying key variables that influence project costs and using mathematical models to calculate estimates based on these variables.

1. Understanding parametric estimating

- **Historical data**: Parametric estimating relies on historical data from past projects to establish relationships between project characteristics and costs.

- **Project characteristics**: Key project characteristics, such as size, scope, complexity, and location, are identified as variables that impact costs.

- **Mathematical models**: Parametric estimating uses mathematical models or algorithms to quantify the relationships between project characteristics and costs.

2. Application of parametric estimating

- **Data collection**: Gather historical data from past projects, including cost data and project characteristics.

- **Variable identification**: Identify the key variables that influence project costs, such as area, volume, quantity, or complexity.

- **Model development**: Develop parametric models or equations that express the relationships between project characteristics and costs.

- **Parameter estimation**: Estimate the parameters of the model based on historical data and statistical analysis.

- **Cost calculation**: Use the parametric model to calculate cost estimates for the current project based on its specific characteristics.

3. Advantages of parametric estimating

- **Accuracy**: Parametric estimating can produce accurate cost estimates by leveraging historical data and empirical relationships between project characteristics and costs.
- **Efficiency**: Once the parametric model is developed, cost estimates can be generated quickly and efficiently for new projects.
- **Flexibility**: Parametric estimating allows for flexibility in adjusting estimates based on changes in project characteristics or input parameters.

4. Limitations of parametric estimating

- **Data availability**: Parametric estimating relies heavily on the availability of accurate and relevant historical data, which may not always be readily accessible.
- **Model assumptions**: The accuracy of parametric estimates depends on the validity of the assumptions underlying the parametric model.
- **Complexity**: Developing parametric models can be complex and requires expertise in statistical analysis and modeling techniques.

5. Best practices for using parametric estimating

- **Data quality**: Ensure the accuracy and reliability of historical data used for developing parametric models.
- **Model validation**: Validate the parametric model against real-world data and expert judgment to assess its accuracy and reliability.

- **Sensitivity analysis**: Conduct sensitivity analysis to assess the impact of changes in input parameters on cost estimates and identify potential sources of uncertainty.

- **Continuous improvement**: Continuously refine and update parametric models based on feedback from actual project costs and evolving industry trends.

Benefits and limitations of each method

1. Unit cost approach

Benefits

- **Simplicity**: Easy to understand and implement, making it suitable for beginners in construction estimating.

- **Speed**: Allows for quick estimation of costs by applying pre-determined unit prices to quantities.

- **Consistency**: Promotes consistency in estimating practices and results across different projects.

Limitations

- **Lack of specificity**: May not account for project-specific factors or variations in material quality, labor productivity, or project conditions.

- **Limited accuracy**: Relies on generalized unit prices, which may not accurately reflect the actual costs for all project components.

- **Dependency on data**: Accuracy of estimates depends on the availability and reliability of unit cost data.

Practical example: Estimating the cost of flooring installation in a residential project using a standard unit price per square foot for the chosen flooring material.

2. Assembly cost estimating

Benefits

- **Efficiency**: Streamlines the estimating process by grouping related items together into pre-defined assemblies.

- **Accuracy**: Provides accurate estimates by considering the collective costs of components within each assembly.

- **Simplicity**: Offers a structured approach to estimating costs, particularly suitable for projects with repetitive elements.

Limitations

- **Limited flexibility**: May not accommodate changes or variations in project requirements outside of pre-defined assemblies.

- **Scope constraints**: Applicability may be limited for projects with unique or custom elements not covered by standard assemblies.

- **Dependency on assemblies**: Accuracy of estimates relies on the availability and accuracy of pre-defined assemblies.

- **Practical example**: Estimating the cost of a wall assembly in a commercial building using a predefined assembly that includes materials, labor, and equipment for wall construction.

3. Parametric estimating

Benefits

- **Accuracy**: Produces accurate estimates by leveraging historical data and empirical relationships between project characteristics and costs.

- **Efficiency**: Allows for quick estimation of costs once the parametric model is developed, reducing time and effort.

- **Flexibility**: Provides flexibility to adjust estimates based on changes in project characteristics or input parameters.

Limitations

- **Data requirements**: Relies heavily on the availability and quality of historical data, which may not always be readily accessible.

- **Model assumptions**: Accuracy of estimates depends on the validity of assumptions underlying the parametric model.

- **Complexity**: Developing parametric models can be complex and requires expertise in statistical analysis and modeling techniques.

Practical example: Estimating the cost of a new residential building based on historical data from similar past projects, considering factors such as size, location, and complexity.

In practice, each cost estimating method has its advantages and limitations, and the choice of method depends on factors such as project size, complexity, available data, and estimator experience. By understanding the benefits and limitations of each method and applying them judiciously, construction estimators can develop accurate and reliable cost estimates tailored to the specific requirements of each project.

CHAPTER 5: LABOR ESTIMATING ESSENTIALS

Labor estimation begins with the identification and classification of construction labor categories, which typically encompass various types of skilled and unskilled labor required to complete a project. By categorizing labor activities based on skill level, specialization, and trade, estimators can develop more accurate and detailed labor cost estimates.

1. Skilled labor categories

Skilled labor categories comprise workers with specialized training, certification, or experience in specific trades or disciplines. These workers possess the expertise and proficiency required to perform complex tasks accurately and efficiently.

- **Examples**: Carpenters, electricians, plumbers, HVAC technicians, masons, welders, painters, and equipment operators.

- **Characteristics**: Skilled laborers typically require formal training, apprenticeship, or certification in their respective trades and may command higher wages due to their specialized skills and expertise.

- **Labor Rates**: Labor rates for skilled workers vary based on factors such as trade, experience, geographic location, union affiliation, and prevailing market conditions.

2. Semi-skilled labor categories

Semi-skilled labor categories encompass workers with intermediate-level skills and experience, capable of performing moderately complex tasks under supervision or guidance.

- **Examples**: Construction laborers, carpenter's assistants, electrician's helpers, pipefitter's assistants, and HVAC technician's assistants.

- **Characteristics**: Semi-skilled laborers typically possess basic knowledge and proficiency in construction tasks but may require supervision or training to perform more advanced duties.

- **Labor rates**: Labor rates for semi-skilled workers are generally lower than those for skilled labor but higher than those for unskilled labor, reflecting their intermediate level of expertise and responsibility.

3. Unskilled labor categories

Unskilled labor categories consist of workers who perform routine, repetitive tasks that require minimal training, expertise, or specialized skills.

- **Examples**: General laborers, construction helpers, material handlers, and site cleanup crew members.

- **Characteristics**: Unskilled laborers typically perform manual labor tasks such as carrying materials, digging trenches, cleaning worksites, and assisting skilled workers.

- **Labor rates**: Labor rates for unskilled workers are typically the lowest among labor categories, reflecting the limited skill requirements and lower entry barriers for these positions.

4. Specialized labor categories

Specialized labor categories encompass workers with expertise in niche or specialized trades or disciplines that require specific skills, certifications, or training beyond standard industry practices.

- **Examples**: Welding inspectors, crane operators, riggers, certified safety professionals, and specialized equipment technicians.

- **Characteristics**: Specialized laborers possess advanced skills and certifications relevant to their specialized roles, often requiring additional training or qualifications beyond standard trade requirements.

- **Labor Rates**: Labor rates for specialized workers vary based on the level of expertise, demand, and scarcity of qualified professionals in their respective fields.

By identifying and classifying construction labor categories based on skill level, specialization, and trade, estimators can develop more accurate labor cost estimates tailored to the specific requirements of each project.

Estimating labor hours and crew sizes for different tasks

Labor estimation involves determining the number of labor hours required to complete specific tasks and establishing appropriate crew sizes based on project requirements, productivity rates, and scheduling constraints.

1. Task analysis

- **Breakdown tasks**: Divide the project into individual tasks or activities, considering factors such as complexity, sequence, and resource requirements.

- **Identify labor requirements**: Determine the labor-intensive tasks that require significant manpower and resources for completion.

2. Labor hour estimation

- **Work breakdown structure**: Develop a work breakdown structure that outlines the tasks, durations, and labor requirements for each activity.

- **Task duration**: Estimate the time required to complete each task based on historical data, industry standards, and project-specific considerations.

- **Consider factors**: Account for factors such as task complexity, skill level required, site conditions, and equipment availability when estimating labor hours.

3. Crew size determination

- **Optimal crew size**: Determine the optimal crew size for each task based on the scope of work, labor availability, productivity rates, and safety considerations.

- **Crew composition**: Consider the mix of skilled, semi-skilled, and unskilled laborers needed for each crew to ensure efficient task execution.

- **Coordination**: Coordinate with project managers, supervisors, and subcontractors to align crew sizes with project schedules and resource allocations.

4. Resource allocation

- **Balanced workloads**: Distribute labor resources evenly across tasks to avoid overloading or underutilizing crews.
- **Schedule alignment**: Align labor hours and crew sizes with project milestones, deadlines, and critical path activities to maintain schedule adherence.
- **Flexibility**: Maintain flexibility to adjust crew sizes and resource allocations based on changing project conditions, unforeseen delays, or resource constraints.

Understanding labor productivity rates

Labor productivity rates quantify the efficiency of labor utilization and provide insights into the effectiveness of workforce management, resource allocation, and project planning.

1. Definition

The measure of the output or work accomplished by a labor crew within a specific timeframe, usually expressed as units of work per labor hour or labor cost per unit of work. Labor productivity rates serve as a key performance indicator for evaluating project performance, identifying inefficiencies, and improving productivity.

2. Factors influencing productivity

- **Skill level**: Skilled laborers typically exhibit higher

productivity rates than semi-skilled or unskilled workers due to their specialized training and expertise.
- **Work environment**: Factors such as weather conditions, site accessibility, material availability, and equipment reliability can impact labor productivity.
- **Management practices**: Effective project management, supervision, communication, and coordination contribute to improved labor productivity.

3. Measurement and analysis

- **Data collection**: Collect data on labor hours, output quantities, and project progress to calculate productivity rates.
- **Performance tracking**: Monitor labor productivity regularly throughout the project lifecycle to identify trends, deviations, and areas for improvement.
- **Benchmarking**: Compare labor productivity rates against industry benchmarks, historical data, or similar projects to assess performance relative to established standards.

4. Continuous improvement

- **Root cause analysis**: Investigate factors contributing to variations in productivity rates and address underlying issues through corrective actions and process improvements.
- **Training and development**: Invest in training programs, skill development initiatives, and technology adoption to enhance labor productivity and efficiency.
- **Feedback mechanisms**: Encourage feedback and input from frontline workers, supervisors, and project teams to identify opportunities for optimizing workflows and enhancing productivity.

Factoring in labor fringe benefits and payroll taxes

Labor cost estimation involves accounting not only for direct wages but also for additional expenses such as fringe benefits and payroll taxes, which are essential components of total labor costs.

1. Fringe benefits

Fringe benefits are non-wage compensations provided to employees in addition to their regular wages, including health insurance, retirement plans, paid time off, and bonuses. Estimate the cost of fringe benefits as a percentage of direct wages, taking into account employer contributions, administrative expenses, and employee eligibility criteria. Comply with legal regulations governing the provision of fringe benefits, including mandatory benefits such as workers' compensation and unemployment insurance.

2. Payroll taxes

- **Types of taxes**: Payroll taxes include federal, state, and local taxes withheld from employee wages to fund government programs such as Social Security, Medicare, income tax, and unemployment insurance.

- **Tax rates**: Determine applicable tax rates based on jurisdictional requirements, employee earnings, and tax brackets established by tax authorities.

- **Employer contributions**: In addition to withholding employee taxes, employers are responsible for making matching contributions for certain taxes, such as Social Security and Medicare.

Staying updated with local labor laws and rates

To ensure accurate labor cost estimation and compliance with regulatory requirements, construction estimators must stay updated with local labor laws, regulations, and prevailing wage rates.

1. Regulatory compliance

- **Minimum wage laws**: Familiarize yourself with federal, state, and local minimum wage laws that establish the minimum hourly wage rates for covered employees.
- **Prevailing wage laws**: Adhere to prevailing wage laws that mandate minimum wage rates for construction projects funded by government contracts or subsidies.
- **Overtime regulations**: Understand overtime regulations governing the payment of overtime wages for hours worked beyond standard workweek or daily limits.

2. Prevailing wage rates

- **Research**: Conduct research to determine prevailing wage rates for construction trades and occupations in your geographic area, typically published by government agencies or labor departments.
- **Applicability**: Apply prevailing wage rates to public works projects, government contracts, and projects subject to prevailing wage requirements to ensure compliance with legal obligations.
- **Documentation**: Maintain accurate records of prevailing wage rates used in labor cost estimates, including their

sources and effective dates, to support transparency and accountability.

3. Industry updates

- **Stay informed**: Stay abreast of industry news, updates, and developments related to labor laws, regulations, and prevailing wage rates through industry publications, professional associations, and government websites.

- **Continuous learning**: Invest in ongoing education and professional development opportunities to deepen your understanding of labor law compliance and stay updated with evolving regulatory requirements.

By factoring in labor fringe benefits and payroll taxes and staying updated with local labor laws and rates, construction estimators can develop accurate labor cost estimates and ensure compliance with legal and regulatory requirements.

CHAPTER 6: MATERIAL ESTIMATING FUNDAMENTALS

Material estimation begins with the identification and classification of construction materials based on their type, function, and usage in the project. By categorizing materials systematically, estimators can develop more accurate material lists and cost estimates.

1. Common construction materials

- **Structural materials**: Materials used for building the structural framework of a project, including concrete, steel, timber, and masonry.

- **Finishing materials**: Materials applied to surfaces for aesthetic or functional purposes, such as drywall, flooring, paint, tiles, and finishes.

- **Mechanical and electrical materials**: Components for mechanical, electrical, and plumbing systems, such as pipes, fittings, conduits, wires, fixtures, and equipment.

- **Site materials**: Materials used in site preparation and landscaping, including earthwork materials, aggregates, soil, gravel, and landscaping elements.

2. Classification criteria

- **Functionality**: Classify materials based on their intended function or purpose in the project, such as structural, architectural, mechanical, or site-related.

- **Durability and lifespan**: Consider the durability, longevity, and maintenance requirements of materials when classifying them for estimation purposes.

- **Availability and sourcing**: Classify materials based on their availability, sourcing options, lead times, and supply chain considerations to ensure timely procurement.

3. Material properties and specifications

- **Physical properties**: Consider the physical characteristics of materials, including dimensions, weight, density, strength, and thermal properties, when estimating quantities.

- **Specifications**: Refer to project specifications, drawings, and standards to identify the required materials, quality standards, grades, sizes, and finishes specified for each application.

4. Specialty and custom materials

- **Specialty materials**: Identify specialty or custom materials required for unique project requirements, specialized applications, or design features.

- **Custom fabrication**: Consider materials that require custom fabrication or manufacturing processes, such as custom millwork, metalwork, or prefabricated components.

5. Waste factors and allowances

- **Waste factors**: Account for material waste factors when estimating quantities to compensate for losses during handling, cutting, installation, and unforeseen circumstances.
- **Allowances**: Include allowances for contingencies, overages, and unforeseen variations in material quantities or specifications during construction.

Understanding material waste factors

Material waste factors refer to the amount of material that is lost or unused during the construction process due to various factors such as handling, cutting, installation, and unforeseen circumstances. Properly accounting for material waste is essential for avoiding cost overruns and delays caused by material shortages.

Material handling and storage

Effective material handling practices can minimize waste by reducing damage and spoilage during transportation, storage, and handling. Proper storage conditions, handling equipment, and protective measures can help prevent material losses due to mishandling or exposure to environmental factors. Material waste often occurs during cutting, shaping, and fabrication processes, particularly for materials such as lumber, steel, and stone. Optimizing cutting patterns, using efficient cutting tools, and minimizing offcuts can help reduce material waste and maximize material utilization.

During installation and construction activities, material waste may result from errors, inaccuracies, rework, and adjustments made to accommodate site conditions or design changes. Proper planning, coordination, and supervision can minimize waste by ensuring accurate measurements, proper fitting, and efficient installation techniques. Despite careful planning and execution, unforeseen circumstances such as design revisions, material defects, errors, and accidents can lead to material waste. Maintaining contingency plans, monitoring project progress, and addressing issues promptly can mitigate the impact of unforeseen waste on project costs and schedules.

To minimize material waste, construction professionals can implement various mitigation strategies such as:

Conducting thorough material takeoffs and estimating quantities accurately. Using prefabricated components and modular construction techniques to reduce onsite fabrication and waste. Implementing lean construction principles to optimize material flows, minimize waste, and improve productivity. Recycling and repurposing waste materials whenever feasible to reduce environmental impact and disposal costs.

Continuous monitoring, analysis, and improvement of material waste management practices are essential for optimizing resource utilization, reducing costs, and enhancing project efficiency. By identifying root causes of waste, implementing corrective actions, and fostering a culture of waste reduction, construction teams can achieve sustainable and cost-effective project outcomes.

Estimating material costs based on quantities and unit prices

Estimating material costs involves calculating the total cost of materials required for a project based on the quantities needed and the unit prices of those materials. This process ensures accurate budgeting and procurement planning.

1. Quantity estimation

- Conduct a thorough material takeoff to determine the quantities of each material required for the project.
- Consider project specifications, drawings, and construction plans to identify material requirements for each component and phase of the project.
- Use standardized measurement units such as square footage, cubic yards, linear feet, or units to quantify material quantities accurately.

2. Unit price determination

- Obtain unit prices for materials from suppliers, vendors, or industry databases.
- Consider factors such as material quality, grade, size, and supplier discounts when determining unit prices.
- Account for variations in unit prices based on quantity discounts, bulk purchasing agreements, and market fluctuations.

3. Cost calculation

- Multiply the quantities of each material by their respective unit prices to calculate the total cost of materials for the project.
- Summarize the costs of all materials to derive the total material cost estimate for the project.
- Consider additional factors such as taxes, surcharges, and handling fees when calculating material costs.

Considering transportation

and delivery costs

Transportation and delivery costs are often overlooked but can significantly impact project budgets and schedules. Estimating these costs accurately ensures timely material delivery and cost-effective logistics management.

1. Distance and location

- Determine the distance between material suppliers and the project site to estimate transportation distances accurately.
- Consider factors such as road conditions, traffic patterns, and geographic terrain when assessing transportation routes and logistics.

2. Mode of transportation

- Evaluate different modes of transportation such as trucking, rail, air, or sea freight based on cost, speed, and suitability for the type of materials being transported.
- Select the most cost-effective and efficient transportation option that meets project requirements and delivery schedules.

3. Delivery logistics

- Coordinate with suppliers, logistics providers, and project stakeholders to plan and schedule material deliveries.
- Anticipate potential delays, bottlenecks, and logistical challenges that may impact delivery timelines and adjust plans accordingly.

4. Cost allocation

- Factor transportation and delivery costs into the overall

material cost estimate by adding them to the cost of materials.

- Allocate transportation costs proportionally to each material based on quantity, weight, or value to ensure accurate cost distribution.

5. Contingency planning

- Include contingency allowances for transportation and delivery costs to account for unforeseen circumstances such as fuel price fluctuations, route changes, or delivery delays.

- Monitor transportation logistics closely and implement contingency plans as needed to mitigate risks and minimize disruptions to the project schedule.

PART 3: REFINING ESTIMATE FOR ACCURACY AND EFFICIENCY

CHAPTER 7: OVERHEAD AND PROFIT CONSIDERATIONS

In this chapter, we consider the essential aspects of overhead costs and profit considerations in construction estimating. Understanding overhead costs and profit margins is crucial for developing accurate and competitive bids while ensuring profitability and sustainability in construction projects.

Types of construction overhead costs

Overhead costs, also known as indirect costs, are expenses incurred in the course of business operations that cannot be directly attributed to a specific project or activity but are necessary for the overall operation of the construction company. Identifying and understanding the various types of overhead costs is essential for accurately allocating them to estimates and determining the true cost of a project.

1. Administrative overhead

Administrative expenses associated with general management,

office operations, and administrative staff salaries.
Includes costs such as office rent, utilities, insurance, office supplies, administrative salaries, and professional services (e.g., accounting, legal).

2. Equipment and tools

Costs associated with the ownership, maintenance, and operation of construction equipment, machinery, and tools.
Includes expenses such as equipment depreciation, maintenance, repairs, fuel, insurance, and equipment rental fees.

3. Project management

Costs related to project management activities, including project supervision, coordination, planning, and administration.
Includes salaries of project managers, site supervisors, project engineers, project management software, communication tools, and project-related travel expenses.

4. Facilities and infrastructure

Expenses associated with maintaining and operating company facilities, yards, warehouses, and storage facilities.
Includes costs such as facility rent or mortgage, utilities, property taxes, facility maintenance, security, and janitorial services.

5. General and corporate overhead

General operating expenses and corporate overhead costs incurred by the construction company as a whole.
Includes costs such as corporate salaries, marketing and advertising, professional memberships, training, research and development, and corporate governance.

6. Contingency and risk management

Provision for unforeseen events, risks, and contingencies that

may impact project costs or schedules.
Includes allowances for project contingencies, insurance premiums, bonding costs, legal fees, and risk management activities.

7. Other indirect costs

Miscellaneous overhead expenses that do not fit into specific categories but are essential for business operations.
Includes costs such as permits and licenses, taxes, employee benefits, travel expenses, subcontractor management, and quality control.

Understanding the types of construction overhead costs allows estimators to accurately allocate overhead expenses to project estimates and determine the true cost of construction projects. In the subsequent sections of this chapter, we'll explore methods for calculating and allocating overhead costs to estimates and factoring in profit margins to develop competitive bids.

Calculating and allocating overhead costs to estimates

Example 1: Percentage of direct costs method

Cost identification

- Identify overhead costs such as administrative expenses, equipment maintenance, and project management salaries.

Cost classification

- Classify overhead costs as indirect expenses not directly attributable to specific projects.

Calculation of overhead rate

- Total Overhead Costs = $100,000
- Total Direct Costs (Labor + Materials) = $500,000

- Overhead Rate = (Total Overhead Costs / Total Direct Costs) × 100%
- Overhead Rate = ($100,000 / $500,000) × 100% = 20%

Allocation to project

- For a project with $50,000 in direct costs (labor + materials):
- Project Overhead Cost = Overhead Rate × Total Direct Costs for Project
- Project Overhead Cost = 20% × $50,000 = $10,000

Example 2: Labor Hours Allocation Method

Cost identification

- Identify overhead costs such as office rent, utilities, and administrative salaries.

Cost classification

- Classify overhead costs as indirect expenses not directly tied to specific projects.

Calculation of overhead rate

- Total Overhead Costs = $80,000
- Total Labor Hours Across Projects = 10,000 hours
- Overhead Rate = Total Overhead Costs / Total Labor Hours
- Overhead Rate = $80,000 / 10,000 hours = $8 per labor hour

Allocation to project

- For a project requiring 500 labor hours:
- Project Overhead Cost = Overhead Rate × Labor Hours for

Project
- Project Overhead Cost = $8/hour × 500 hours = $4,000

Example 3: Square Footage Allocation Method

Cost identification
- Identify overhead costs such as facility maintenance, security, and property taxes.

Cost classification
- Classify overhead costs as indirect expenses applicable across all projects.

Calculation of overhead rate
- Total Overhead Costs = $120,000
- Total Square Footage of Facilities = 10,000 sq. ft.
- Overhead Rate = Total Overhead Costs / Total Square Footage
- Overhead Rate = $120,000 / 10,000 sq. ft. = $12 per sq. ft.

Allocation to project
- For a project involving 2,000 sq. ft. of construction:
- Project Overhead Cost = Overhead Rate × Square Footage for Project
- Project Overhead Cost = $12/sq. ft. × 2,000 sq. ft. = $24,000

Factoring in profit margin for a competitive bid

Example 1: Residential construction project

Project details
- Residential house construction project with a total

- estimated cost of $500,000.
- Direct project costs (labor and materials): $400,000.
- Overhead costs allocated to the project: $50,000.

Profit margin calculation

- Gross Profit = Total Project Revenue - Direct Project Costs
- = $500,000 - $400,000
- = $100,000
- Net Profit = Gross Profit - Overhead Costs
- = $100,000 - $50,000
- = $50,000
- Profit Margin (%) = (Net Profit / Total Project Cost) × 100
- = ($50,000 / $500,000) × 100
- = 10%

Competitive bid price

To submit a competitive bid, the construction company may decide to aim for a profit margin of 10%.

- Bid Price = Total Project Cost + Profit Margin
- = $500,000 + ($500,000 × 10%)
- = $500,000 + $50,000
- = $550,000

Example 2: Commercial renovation project

Project details

- Commercial office renovation project with a total estimated cost of $1,200,000.
- Direct project costs (labor and materials): $900,000.
- Overhead costs allocated to the project: $120,000.

Profit margin calculation

- Gross Profit = Total Project Revenue - Direct Project Costs

- = $1,200,000 - $900,000
- = $300,000
- Net Profit = Gross Profit - Overhead Costs
- = $300,000 - $120,000
- = $180,000
- Profit Margin (%) = (Net Profit / Total Project Cost) × 100
- = ($180,000 / $1,200,000) × 100
- = 15%

Competitive bid price

To submit a competitive bid, the construction company may aim for a profit margin of 15%.

- Bid Price = Total Project Cost + Profit Margin
- = $1,200,000 + ($1,200,000 × 15%)
- = $1,200,000 + $180,000
- = $1,380,000

Example 3: Infrastructure development project

Project details

- Infrastructure development project with a total estimated cost of $5,000,000.
- Direct project costs (labor and materials): $4,000,000.
- Overhead costs allocated to the project: $400,000.

Profit margin calculation

- Gross Profit = Total Project Revenue - Direct Project Costs
- = $5,000,000 - $4,000,000
- = $1,000,000
- Net Profit = Gross Profit - Overhead Costs
- = $1,000,000 - $400,000
- = $600,000
- Profit Margin (%) = (Net Profit / Total Project Cost) × 100
- = ($600,000 / $5,000,000) × 100

- = 12%

Competitive bid price

- To submit a competitive bid, the construction company may aim for a profit margin of 12%.
- Bid Price = Total Project Cost + Profit Margin
- = $5,000,000 + ($5,000,000 × 12%)
- = $5,000,000 + $600,000
- = $5,600,000

Industry standards or benchmarks for profit margins

Profit margins in the construction industry can vary widely based on factors such as project type, size, complexity, market conditions, and geographic location. Some general benchmarks and considerations for profit margins in different sectors of the construction industry include the following:

1. Residential construction

- **Custom home building**: Profit margins typically range from 15% to 25%. Custom home projects often involve higher margins due to the personalized nature of the work and higher client expectations.

- **Speculative home building**: Profit margins generally range from 10% to 20%. Builders of speculative homes (built without a specific buyer) often face more market risks, which can impact margins.

2. Commercial construction

- **Office buildings and retail spaces**: Profit margins typically range from 8% to 15%. These projects can vary in complexity and scale, affecting the achievable margins.

- **Industrial facilities**: Profit margins generally range from 5% to 12%. The specialized nature and larger scale of industrial projects often lead to lower profit margins compared to other commercial construction.

3. Infrastructure and civil engineering

- **Roads and bridges**: Profit margins typically range from 5% to 10%. These projects are often highly competitive and subject to strict regulatory and safety standards, impacting margins.

- **Utilities and water treatment plants**: Profit margins generally range from 6% to 12%. The complexity and long-term nature of these projects can influence profit expectations.

4. Renovation and remodeling

- **Residential renovations**: Profit margins typically range from 10% to 20%. Renovation projects can involve higher margins due to the unpredictability and customization required.

- **Commercial remodeling**: Profit margins generally range from 8% to 15%. Commercial remodeling projects can vary significantly, affecting profit margins.

5. Specialty construction

- **Green building and sustainable construction**: Profit margins can range from 10% to 20% or higher. The growing demand for sustainable construction practices can lead to higher margins due to the specialized knowledge and materials required.

- **Historical restoration**: Profit margins typically range from 10% to 25%. The intricate and detailed work involved in historical restoration can justify higher profit margins.

6. Factors influencing profit margins

- **Project complexity and risk**: More complex and riskier projects typically command higher profit margins to compensate for potential challenges and uncertainties.
- **Market conditions**: Economic conditions, competition, and demand in the local market can significantly impact achievable profit margins.
- **Client relationships**: Strong relationships with clients and effective negotiation strategies can influence profit margins positively.
- **Company efficiency**: Operational efficiency, cost control measures, and effective project management can enhance profit margins.

7. Continuous monitoring and adjustment

- Construction companies should regularly review and analyze profit margins on completed projects to assess performance and identify areas for improvement.
- Adjust profit margin targets and bidding strategies based on lessons learned, market changes, and evolving industry standards to maintain competitiveness and profitability.

CHAPTER 8: RISK MANAGEMENT AND CONTINGENCY PLANNING

1. Understanding project risks

Project risks refer to any factors that can potentially impact the scope, schedule, cost, or quality of a construction project. These risks can stem from various sources and need to be proactively managed.

2. Common types of project risks

- **Technical risks**: Issues related to design flaws, technology failures, or construction methods.
- **Financial risks**: Budget overruns, unexpected cost increases, or funding shortfalls.
- **Operational risks**: Delays in material delivery, equipment breakdowns, or labor shortages.
- **Environmental risks**: Weather conditions, natural disasters, or environmental regulations.
- **Legal and regulatory risks**: Compliance with laws,

permits, and contractual obligations.

- **Market risks**: Fluctuations in material costs, labor rates, or economic downturns.
- **Human risks**: Errors, accidents, or personnel changes.

3. Risk identification process

- **Brainstorming sessions**: Conduct brainstorming sessions with the project team to identify potential risks based on past experiences and project specifics.
- **SWOT analysis**: Perform a SWOT analysis (Strengths, Weaknesses, Opportunities, Threats) to identify internal and external risks.
- **Historical data review**: Analyze data from previous projects to identify recurring risks and patterns.
- **Expert consultation**: Seek input from industry experts, consultants, or stakeholders to identify less obvious risks.

4. Documenting identified risks

- Create a risk register to document identified risks, their potential impact, and the likelihood of occurrence.
- Categorize risks based on their source (technical, financial, operational, etc.) for better management.

5. Assessing and prioritizing risks

- **Qualitative assessment**: Evaluate the impact and likelihood of each risk using qualitative measures (e.g., high, medium, low).

- **Quantitative assessment**: Use quantitative methods (e.g., statistical analysis, Monte Carlo simulations) to estimate the potential cost and schedule impacts.
- **Risk matrix**: Develop a risk matrix to prioritize risks based on their severity and probability, focusing on high-impact and high-likelihood risks.

Developing contingency plans to manage cost overruns

1. Understanding contingency planning

Contingency planning involves preparing for unforeseen events and potential cost overruns by setting aside additional resources (time, money, materials) to address these issues if they arise.

2. Establishing contingency reserves

- **Cost contingency**: Allocate a percentage of the total project budget as a cost contingency reserve to cover unexpected expenses. This percentage typically ranges from 5% to 15% depending on the project's complexity and risk profile.

Schedule contingency: Include buffer time in the project schedule to accommodate delays caused by unforeseen events.

3. Types of contingency plans

- **Risk mitigation plans**: Develop strategies to reduce the likelihood or impact of identified risks (e.g., adopting alternative construction methods, securing backup suppliers).

- **Risk transfer plans**: Transfer risks to third parties through insurance policies, warranties, or subcontracting agreements.

- **Risk acceptance plans**: Acknowledge certain risks that cannot be mitigated or transferred and prepare to manage their impacts if they occur.

4. Developing specific contingency actions

- For each high-priority risk, develop specific contingency actions detailing what will be done if the risk materializes.

- Assign responsibilities to team members for implementing these actions and ensure they are included in the project plan.

5. Monitoring and updating contingency plans

- Continuously monitor project progress and risk indicators to identify new risks or changes in existing risks.

- Regularly update contingency plans and reserves based on project developments and feedback from the project team.

6. Communication and training

- Ensure that all stakeholders are aware of the contingency plans and understand their roles in executing them.

- Provide training to the project team on risk management and contingency planning processes.

Including contingency reserves in estimate

1. Importance of contingency reserves

Contingency reserves are crucial for covering unforeseen costs that arise during a construction project. These reserves act as a financial buffer, ensuring that the project can proceed without significant disruptions when unexpected expenses occur.

2. Determining the amount of contingency reserves

- The amount of contingency reserves varies based on project size, complexity, and risk profile. A common approach is to allocate a percentage of the total project budget, typically ranging from 5% to 15%.

- High-risk projects or those with significant uncertainties may require higher contingency reserves.

3. Types of contingency reserves

- **Cost contingency**: Set aside funds specifically for unexpected costs related to labor, materials, equipment, and other project expenses.

- **Schedule contingency**: Include additional time in the project schedule to account for potential delays due to unforeseen events.

4. Incorporating contingency reserves into estimates

- **Initial estimate**: When preparing the initial estimate, include a line item for contingency reserves. Clearly identify this in the estimate to ensure transparency with stakeholders.

- **Budget revisions**: As the project progresses and more information becomes available, adjust the contingency reserves accordingly. Reallocate funds if certain risks are mitigated or new risks are identified.

5. Communicating contingency reserves

- Clearly communicate the rationale for contingency reserves to clients and stakeholders. Explain that these reserves are a standard practice to manage potential risks and ensure project success.

- Transparency about contingency reserves helps build trust and sets realistic expectations for project costs.

Real-world examples of risk management

Example 1: Commercial office building

Project details

- Construction of a new commercial office building in an urban area.
- Estimated project cost: $10,000,000.
- Contingency reserve: 10% ($1,000,000).

Risk identification

- Potential risks included regulatory changes, site conditions, and material price fluctuations.

Contingency planning

- Developed specific plans for regulatory delays, including alternate approval pathways and expedited permitting services.
- Conducted thorough site investigations to mitigate risks related to unknown site conditions.
- Established contracts with material suppliers that included price adjustment clauses to manage material cost fluctuations.

Risk management

- Midway through the project, a new zoning regulation required design modifications. The contingency reserve covered additional design and permitting costs without impacting the overall project budget.
- Unexpected soil contamination was discovered during excavation. The contingency reserve funded the remediation efforts, preventing significant delays and cost overruns.

Example 2: Infrastructure development project

Project details

- Development of a new highway segment.
- Estimated project cost: $50,000,000.
- Contingency reserve: 8% ($4,000,000).

Risk identification

- Key risks included adverse weather conditions, supply chain disruptions, and environmental regulations.

Contingency planning

- Implemented weather monitoring systems and developed plans for weather-related delays.
- Established relationships with multiple suppliers to mitigate the risk of supply chain disruptions.
- Conducted comprehensive environmental assessments and developed mitigation plans to comply with environmental regulations.

Risk management

- A severe weather event caused significant delays in the project schedule. The schedule contingency allowed the project team to extend deadlines without incurring penalties.
- A major supplier experienced a disruption, but pre-arranged agreements with secondary suppliers ensured that material availability was maintained, avoiding costly delays.

Example 3: Residential renovation project

Project details

- High-end residential renovation in a historic district.
- Estimated project cost: $1,500,000.
- Contingency reserve: 12% ($180,000).

Risk identification

- Identified risks included discovering hidden structural issues, delays in obtaining specialized materials, and unexpected client changes.

Contingency planning

- Conducted a thorough pre-renovation inspection to identify potential structural issues.
- Sourced multiple suppliers for specialized materials and pre-ordered critical components.
- Established a clear change order process with the client to manage scope changes.

Risk management

During the renovation, previously hidden structural damage was uncovered. The cost contingency covered the additional structural repairs, ensuring the project stayed on track.

Delays in the delivery of custom-made fixtures were mitigated by the contingency reserve, allowing for temporary solutions and avoiding project downtime.

CHAPTER 9: PRESENTING AND COMMUNICATING ESTIMATE EFFECTIVELY

Creating professional and clear estimate reports

1. Purpose of an estimate report

An estimate report serves as a comprehensive document that outlines the anticipated costs associated with a construction project. It provides clients and stakeholders with a clear understanding of the financial requirements and helps set realistic expectations.

2. Key components of a professional estimate report

a. Cover page

- **Title**: Clearly state that this is an estimate report and include the project name.

- **Company information**: Include your company's name, logo, contact information, and the date of the report.
- **Client information**: Provide the client's name and contact details.

b. Executive summary

- Provide a brief overview of the project, including its scope, objectives, and estimated total cost.
- Highlight key points, such as major cost drivers and critical assumptions.

c. Table of contents

- List all sections and subsections of the report for easy navigation.

d. Project scope

- Define the scope of the project, including a detailed description of the work to be performed.
- Specify inclusions and exclusions to avoid misunderstandings.

e. Detailed cost breakdown

- **Labor costs**: Break down labor costs by category, including hours, rates, and total costs for each task.
- **Material costs**: Itemize materials, including quantities, unit prices, and total costs.
- **Equipment costs**: List any equipment needed, rental rates, and associated costs.
- **Subcontractor costs**: Provide details of subcontractor

services, including their bids and total costs.

- **Overhead and profit**: Clearly state the overhead percentage and profit margin applied to the estimate.

f. Contingency reserves

- Include a section for contingency reserves, explaining the rationale for the percentage or amount allocated.

g. Assumptions and exclusions

- Outline any assumptions made during the estimating process.
- Clearly state what is excluded from the estimate to manage client expectations.

h. Project schedule

- Provide a high-level project timeline, highlighting key milestones and deliverables.
- Indicate the estimated start and completion dates.

i. Risk management

- Summarize potential project risks and the contingency plans in place to address them.

j. Appendices

- Include supporting documents such as detailed takeoff sheets, supplier quotes, and subcontractor bids.
- Provide any additional information that supports the estimate, such as reference projects or certifications.
- Steps to Ensure Clarity and Professionalism

1. Use clear and concise language

- Avoid jargon and technical terms that may confuse the client. Use plain language to explain costs and scope.

2. Consistent formatting

- Use a consistent font, size, and style throughout the document.
- Employ bullet points and numbered lists to organize information logically.

3. Visual aids

- Incorporate charts, graphs, and tables to present data visually. This can help clients quickly grasp cost distributions and key figures.

4. Review and proofread

- Thoroughly review the report for accuracy and completeness. Check for any errors or inconsistencies.
- Have a colleague or supervisor review the report to catch any overlooked details.

5. Digital tools

- Use estimating software and word processing tools to streamline the report creation process and ensure a polished final product.
- Consider using templates to maintain consistency across all estimate reports.

6. Client-centric approach

- Tailor the report to the client's needs and preferences. Customize sections to address specific client concerns or

project requirements.
- Be prepared to explain and discuss the estimate in detail with the client, providing clarity on any questions they may have.

Effectively communicating estimates to stakeholders

Effective communication of construction estimates to stakeholders is crucial for building trust, ensuring transparency, and securing project approval. Stakeholders, including clients, project managers, contractors, and financial backers, need to understand the estimate's rationale, details, and implications to make informed decisions.

Understanding your audience

Identify stakeholders

Clients: Focus on cost, scope, and project outcomes.

Project managers: Emphasize timelines, resource allocation, and risk management.

Contractors: Highlight material specifications, labor requirements, and subcontractor involvement.

Financial backers: Concentrate on return on investment, budget adherence, and financial risks.

Tailor your message

- Customize your communication to address the specific interests and concerns of each stakeholder group. This approach ensures that the information is relevant and easily digestible.

Preparing for the presentation

Organize your data

- Ensure that all data is accurate, up-to-date, and logically organized. This includes cost breakdowns, assumptions, and contingency plans.

Create visual aids

- Use charts, graphs, and tables to visually represent data. Visual aids help stakeholders quickly grasp complex information and see the big picture.

Develop clear handouts

- Prepare concise handouts summarizing key points. These should be easy to read and reference, providing stakeholders with a takeaway that reinforces your presentation.

Rehearse your presentation

- Practice your presentation multiple times to ensure clarity, confidence, and smooth delivery. Anticipate potential questions and prepare responses.

During the presentation

- Start with an Executive Summary
- Begin with a high-level overview of the project, including the total estimated cost, project scope, and key assumptions. This sets the stage for the detailed information to follow.
- Explain the Methodology
- Clearly describe the estimating methods used (e.g., unit cost, assembly cost, parametric estimating) and why they were chosen. This builds credibility and helps

stakeholders understand the basis of your estimates.

Detailed cost breakdown

- Walk through the detailed cost breakdown, covering labor, materials, equipment, subcontractors, overhead, and profit margins. Highlight any significant cost drivers or unique considerations.

Highlight contingency reserves

- Explain the contingency reserves included in the estimate, detailing the percentage or amount allocated and the rationale behind it. This demonstrates your proactive approach to risk management.

Address risks and assumptions

- Discuss identified project risks, their potential impact on costs and timelines, and the contingency plans in place. Also, clarify any assumptions made during the estimating process to manage expectations.

Engage with visual aids

- Use your prepared charts, graphs, and tables to reinforce key points. Visual aids can help stakeholders understand cost distributions and other complex data more effectively.

Invite questions and discussion

- Encourage questions and open discussion to address any concerns or uncertainties. Be prepared to provide detailed explanations and additional information as needed.

After the presentation

- Follow Up with Documentation:
- Provide stakeholders with a copy of the estimate report, handouts, and any other relevant documents. This allows them to review the information at their own pace.

Clarify and confirm

- Follow up with individual stakeholders to clarify any points of confusion and confirm their understanding and acceptance of the estimate.

Regular updates

- Maintain ongoing communication with stakeholders throughout the project. Provide regular updates on cost tracking, budget adherence, and any changes to the estimate.

Feedback loop

- Establish a feedback loop to continuously improve your estimating and communication processes. Solicit feedback from stakeholders on the clarity, completeness, and usefulness of your estimate presentations.

- **Case study**: Effective Communication in Action
- **Project**: Construction of a Mixed-Use Development

Stakeholders involved

- Clients, project managers, contractors, and financial backers.

Communication strategy

- A comprehensive presentation was prepared, starting with an executive summary and followed by a detailed cost breakdown. Visual aids, including pie charts and Gantt charts, were used to illustrate cost allocations and project timelines.

Key points addressed

- The rationale behind the chosen estimating methods was explained, along with the allocation of contingency reserves and risk management plans. Stakeholders were encouraged to ask questions, leading to an engaging and informative discussion.

Outcome

The clear and thorough communication of the estimate helped secure stakeholder approval and funding. Ongoing updates and transparent communication maintained stakeholder confidence throughout the project.

PART 4: MASTERING EFFICIENCY AND TECHNOLOGY IN THE MODERN ERA

CHAPTER 10: UNDERSTANDING CONSTRUCTION ESTIMATING SOFTWARE

Understanding construction estimating software is vital for modern construction professionals as it revolutionizes the estimating process. This software serves as a powerful tool for calculating project costs efficiently and accurately. It operates by automating complex calculations, significantly reducing the risk of human error. Moreover, it centralizes data related to costs and other project details, ensuring consistency across estimates.

One of the primary advantages of construction estimating software is its ability to enhance efficiency. By automating processes that would otherwise be time-consuming, it speeds up the overall estimating process. Additionally, it offers pre-built templates and material libraries, simplifying adjustments and updates.

Another key aspect of understanding construction estimating software is its role in organizing project information. By centralizing documents, drawings, and data, it provides easy access and management. This organization is crucial for maintaining version control and ensuring that all stakeholders are working with the most up-to-date information.

Despite its benefits, there are considerations to bear in mind when utilizing construction estimating software. Implementing such software can be costly, requiring significant upfront investment. Additionally, extensive training may be necessary for users to navigate the software effectively, particularly for more complex systems.

The accuracy of estimates produced by construction estimating software relies heavily on the quality and accuracy of input data. As such, regular updates and maintenance of the database are necessary to ensure the reliability of the estimates. Furthermore, some software solutions may have a steep learning curve, potentially hindering productivity during the initial stages of implementation.

Despite these challenges, understanding construction estimating software empowers construction professionals to make informed decisions about its integration into their workflows. By leveraging its capabilities effectively, they can streamline the estimating process and enhance overall project efficiency.

Benefits and limitations of construction estimating software

Construction estimating software has become an invaluable tool for project managers, estimators, and contractors, enhancing the accuracy, efficiency, and organization of the estimating process. These software solutions perform complex calculations quickly and accurately, reducing the risk of human error. By utilizing a centralized database for costs and other data, the software ensures consistency across estimates.

Automated processes save time compared to manual calculations and data entry, speeding up the overall estimating process. Pre-built templates and material libraries streamline the process further, allowing for quick adjustments and updates. Additionally, the software organizes all documents, drawings, and data in one place, making it easy to access and manage project information. This centralization aids in version control, ensuring that users always work with the most up-to-date information.

Access to real-time pricing and cost data helps create more accurate and current estimates, aiding in better cost management. Estimating software also facilitates the monitoring of costs throughout the project lifecycle, aiding in budget adherence and financial planning. Many estimating software solutions integrate with project management tools, promoting seamless transitions from estimating to project execution. Moreover, cloud-based solutions allow for remote access and real-time collaboration among team members.

However, there are some limitations to consider. Implementing estimating software can be expensive, requiring significant upfront costs. Furthermore, extensive training may be necessary to familiarize staff with the software, particularly for complex systems. The accuracy of estimates heavily depends on the quality and accuracy of input data. Inaccurate or outdated data can lead to incorrect estimates, necessitating regular updates and maintenance of the database.

Some estimating software solutions can be complex and difficult to navigate, especially for users who are not tech-savvy. A steep learning curve may hinder productivity initially as users become familiar with the software. Over-reliance on software can also be a risk, as human oversight is still necessary to ensure estimates are realistic and consider unique project conditions. Additionally, not all estimating software integrates seamlessly with other tools and systems, potentially causing compatibility issues.

Types of estimating software

Two main categories of estimating software are commonly used: cloud-based and desktop applications. Each type offers distinct advantages and considerations, catering to different user preferences and project requirements.

1. Cloud-based estimating software

Cloud-based estimating software operates on remote servers accessed through the internet. Users can access the software from any device with an internet connection, providing flexibility and mobility.

Advantages

- **Accessibility**: Users can access the software from anywhere, using any device with internet access.
- **Collaboration**: Enables real-time collaboration among team members, allowing multiple users to work on the same estimate simultaneously.
- **Automatic updates**: Software updates and maintenance are handled by the provider, ensuring users always have access to the latest features and improvements.
- **Scalability**: Cloud-based solutions often offer scalable pricing plans, allowing users to adjust their subscription based on their needs.

Considerations

- **Internet dependence**: Relies on a stable internet connection for access, which may pose challenges in areas with poor connectivity.
- **Security concerns**: Data is stored on external servers, raising potential security and privacy concerns.

Subscription costs: Typically involves a subscription-based pricing model, which may result in ongoing costs.

2. Desktop estimating software

Desktop estimating software is installed and operated locally on a user's computer or network. It does not require an internet connection for use, offering greater control over data storage and security.

Advantages

- **Offline access**: Users can work offline without internet access, making it suitable for remote job sites or areas with limited connectivity.

- **Data control**: Data is stored locally, giving users greater control over security and privacy.

- **One-time purchase**: Often involves a one-time purchase fee rather than ongoing subscription costs.

Considerations

- **Limited accessibility**: Restricted to the device on which the software is installed, limiting accessibility compared to cloud-based solutions.

- **Manual updates**: Users are responsible for manually updating the software, which may require additional time and effort.

- **Collaboration challenges**: Collaboration among team members may be more challenging compared to cloud-based solutions, as files need to be shared manually.

Comparison chart or detailed review of popular software

options

Cloud-based estimating software

ProEst

- **Features**: Comprehensive estimating tools, integration with project management software, cloud-based collaboration.
- **Pricing**: Subscription-based pricing model.
- **Pros**: Robust feature set, real-time collaboration capabilities.
- **Cons**: Higher subscription costs.

Clear estimates

- **Features**: User-friendly interface, customizable templates, online proposal generation.
- **Pricing**: Tiered subscription plans based on company size.
- **Pros**: Intuitive design, customizable templates.
- **Cons**: Limited advanced features compared to other options.

Desktop estimating software

HeavyBid

- **Features**: Advanced estimating tools, integration with accounting systems, offline access.
- **Pricing**: One-time purchase with optional maintenance plan.
- **Pros**: Powerful feature set, offline access.
- **Cons**: Higher upfront costs, limited collaboration features.

RSMeans data online

- **Features**: Extensive cost data library, integration with estimating software, offline access with data synchronization.
- **Pricing**: Subscription-based pricing model.
- **Pros**: Rich database of cost information, integration options.
- **Cons**: Subscription costs may add up over time.

PART 5: PUTTING IT ALL TOGETHER: PRACTICAL APPLICATIONS

CHAPTER 11: BIDDING AND PROJECT MANAGEMENT INTEGRATION

Preparing a bid package for construction projects is a meticulous process that requires careful attention to detail and thorough documentation. This package serves as the foundation for contractors to submit competitive bids and secure projects. Typically, a bid package includes a variety of documents and information essential for both the contractor and the client.

The bid package should include a detailed project scope outlining the specific requirements, objectives, and deliverables of the construction project. This serves as a reference point for contractors to understand the scope of work and tailor their bids accordingly. Additionally, the package should provide any relevant drawings, plans, and specifications to help contractors accurately assess the project requirements.

In addition to the project scope, the bid package should include clear instructions on how contractors should submit their bids. This may include details on the format and structure of the bid proposal, required documents, and deadlines for submission. Providing clear guidance ensures that all bids are submitted in a consistent and organized manner, facilitating the evaluation process.

The bid package should also specify any contractual terms and conditions that contractors must adhere to when submitting their bids. This may include information on insurance requirements, bonding requirements, payment terms, and other contractual obligations. Clearly outlining these terms helps mitigate misunderstandings and ensures that all parties are aware of their responsibilities.

Another crucial component of the bid package is the cost breakdown or estimate for the project. This should detail the anticipated costs associated with the project, including labor, materials, equipment, overhead, and profit margin. Providing a comprehensive cost breakdown allows contractors to accurately assess the project's financial requirements and develop competitive bids.

Lastly, the bid package may include any additional information or requirements specific to the project or client. This could include references, qualifications, past performance evaluations, or any other criteria used to evaluate bids. Tailoring the bid package to the unique needs of the project helps ensure that contractors provide relevant information and proposals.

Key elements of a winning bid

- **Understanding client needs**: Thoroughly analyze project requirements and tailor solutions accordingly.
- **Showcasing expertise**: Highlight relevant experience, qualifications, and past successes.
- **Transparent cost breakdown**: Provide detailed and competitive estimates for labor, materials, and

overhead.

- **Effective communication**: Maintain professionalism and responsiveness throughout the bidding process.
- **Innovative solutions**: Offer creative and value-added services aligned with the project goals.
- **Reliability and commitment**: Assure timely completion, adherence to budget, and quality standards.
- **Professional presentation**: Present the bid in a clear, concise, and visually appealing manner.

Submitting and following up on bids

Submitting and following up on bids is a crucial process in the construction industry, where securing projects often hinges on effective communication and attention to detail. Once a bid is accurately prepared, it must be submitted in adherence to the client's specifications and deadlines. This entails compiling all required documentation, including detailed cost estimates, project plans, and relevant qualifications, and submitting them via the designated method stipulated by the client.

Following the submission, it is imperative to maintain proactive communication with the client to ensure the bid has been received and to address any queries or concerns they may have promptly. This follow-up communication not only demonstrates professionalism but also provides an opportunity to reinforce key aspects of the bid and emphasize the contractor's commitment to the project.

Staying engaged throughout the bid evaluation process is essential. By actively monitoring the progress and promptly responding to any requests for clarification or additional information, contractors can demonstrate their responsiveness and dedication to meeting the client's needs.

Using estimates in project planning and cost control

Utilizing estimates in project planning and cost control is integral to ensuring the successful execution of construction projects within budgetary constraints and timelines. Estimates serve as foundational tools that guide project planning from inception to completion, facilitating informed decision-making and effective resource allocation.

In the initial stages of project planning, estimates provide essential insights into the anticipated costs associated with various aspects of the project, including materials, labor, equipment, and overhead. By analyzing these estimates, project managers can develop comprehensive budgets, establish realistic timelines, and identify potential risks or challenges that may impact project delivery.

As the project progresses, estimates serve as benchmarks against which actual costs and progress are measured. Through ongoing cost tracking and comparison with initial estimates, project managers can identify deviations from the planned budget and implement timely adjustments to mitigate cost overruns or delays. This proactive approach to cost control enables stakeholders to make informed decisions and take corrective actions as needed to keep the project on track.

It plays a crucial role in optimizing resource allocation and prioritizing project activities. By aligning estimated costs with project objectives and priorities, project managers can allocate resources efficiently, ensuring that critical tasks are completed on time and within budget. This strategic allocation of resources minimizes waste, maximizes productivity, and enhances overall project performance.

Estimates also facilitates effective communication and

collaboration among project stakeholders. By providing transparent and reliable cost projections, estimates foster a shared understanding of project goals, constraints, and expectations. This collaborative approach fosters trust and accountability among team members, leading to enhanced coordination and synergy throughout the project lifecycle.

Tracking and adjusting estimates during the project

Tracking and adjusting estimates during the project is an essential aspect of effective project management, allowing for real-time assessment and adaptation to changing circumstances. This process involves ongoing monitoring of project progress and costs against initial estimates, identifying variances, and making necessary adjustments to ensure project objectives are met within budgetary constraints.

- **Continuous monitoring**: Project managers regularly monitor project progress and costs against initial estimates to identify any discrepancies or deviations that may arise during the execution phase.

- **Identifying variances**: Discrepancies between estimated and actual costs, timelines, or resource utilization are identified through regular tracking and analysis. These variances may result from factors such as changes in scope, unexpected delays, or fluctuations in material or labor costs.

- **Root cause analysis**: Once variances are identified, project managers conduct root cause analysis to determine the underlying factors contributing to the discrepancies. This may involve reviewing project documentation, conducting stakeholder interviews, or analyzing historical data.

- **Adjusting estimates**: Based on the findings of the

root cause analysis, project managers adjust estimates as necessary to reflect the current project status and forecasted outcomes. This may involve revising cost projections, updating timelines, or reallocating resources to address emerging challenges.
- **Communicating changes**: Project managers communicate any adjustments to estimates with relevant stakeholders, including clients, team members, and vendors. Clear and transparent communication ensures that all parties are informed of changes and aligned with the revised project plan.
- **Revising project plans**: Adjustments to estimates may necessitate revisions to the overall project plan, including timelines, milestones, and deliverables. Project managers update project documentation and communicate revised plans to stakeholders to ensure alignment and clarity.
- **Iterative process**: Tracking and adjusting estimates is an iterative process that continues throughout the project lifecycle. Project managers regularly review and update estimates based on evolving project dynamics, ensuring that the project remains on track and within budget.

By tracking and adjusting estimates during the project, project managers can proactively manage project risks, optimize resource utilization, and maintain project alignment with client expectations. This iterative approach to estimating fosters agility and adaptability, enabling projects to respond effectively to changing conditions and deliver successful outcomes.

Post-bid activities and how estimates evolve during project execution

Post-bid activities encompass a series of crucial tasks and processes that occur after the bid has been submitted and awarded. These activities play a vital role in ensuring the successful execution of the project and include contract negotiation, project kickoff, resource allocation, and project planning. As the project progresses, estimates evolve dynamically in response to changes in scope, unforeseen challenges, and emerging opportunities. This evolution reflects the iterative nature of project management and requires continuous monitoring, analysis, and adjustment to ensure project success.

- **Contract negotiation**: Following bid acceptance, contractors engage in contract negotiation with the client to finalize terms, conditions, and deliverables. This phase involves clarifying project requirements, establishing milestones, and confirming budgetary allocations.
- **Project kickoff**: Once the contract is finalized, the project kickoff marks the commencement of project execution. During this phase, project teams are assembled, roles and responsibilities are assigned, and project objectives are reaffirmed.
- **Resource allocation**: Project managers allocate resources based on the project scope, timeline, and budget. This includes assigning personnel, equipment, and materials to specific tasks and activities as outlined in the project plan.
- **Project planning**: Project planning continues post-bid, with a focus on refining and detailing the project schedule, budget, and deliverables. This involves developing work breakdown structures, establishing task dependencies, and identifying critical path activities.

As the project progresses, estimates evolve in response to changing conditions and project dynamics. Several factors contribute to this evolution:

- **Scope changes**: Adjustments to the project scope, whether due to client requests, regulatory requirements, or unforeseen circumstances, impact estimates for both costs and timelines.

- **Risk management**: As project risks materialize or new risks emerge, estimates may need to be adjusted to account for potential impacts on project costs and schedules.

- **Resource utilization**: Variations in resource availability, productivity rates, or material costs may necessitate revisions to estimates to reflect actual project conditions.

- **Change orders**: Change orders initiated by the client or as a result of project changes require modifications to estimates to incorporate additional costs or schedule adjustments.
- **Performance tracking**: Ongoing monitoring of project performance against planned targets provides valuable insights into project progress and allows for the refinement of estimates based on actual data.

Post-bid activities involve contract negotiation, project kickoff, resource allocation, and project planning, all of which contribute to the successful execution of the project. Estimates evolve dynamically during project execution in response to changes in scope, risk, resource utilization, and performance tracking, requiring continuous monitoring, analysis, and adjustment to ensure project success.

PART 6: BEYOND THE BASICS

CHAPTER 12: ADVANCED ESTIMATING TECHNIQUES

Value engineering is a systematic approach to project management that aims to maximize the value of a project while minimizing costs. It involves analyzing various aspects of a project to identify opportunities for cost savings without compromising quality or functionality. By scrutinizing every element of the project, from materials and design to construction methods and processes, value engineering seeks to optimize efficiency and resource utilization.

One of the primary objectives of value engineering is to identify and eliminate unnecessary costs without sacrificing project objectives or performance. This may involve reevaluating design specifications, exploring alternative materials or construction methods, or streamlining processes to achieve the same results at a lower cost. Value engineering encourages creative problem-solving and innovation, challenging project stakeholders to think outside the box and find innovative solutions to complex challenges.

A key aspect of value engineering is collaboration among project stakeholders, including architects, engineers, contractors, and clients. By bringing together diverse perspectives and expertise, value engineering workshops or sessions facilitate brainstorming and idea generation, leading to the identification of cost-saving opportunities that may not have been apparent otherwise. Through open communication and collaboration, stakeholders can collectively evaluate potential changes and determine the most effective strategies for achieving cost savings without compromising project quality or performance.

Value engineering is not a one-time exercise but rather an ongoing process that should be integrated into the project lifecycle from the outset. By incorporating value engineering principles early in the design phase, project teams can proactively identify cost-saving opportunities and make informed decisions that have a significant impact on project outcomes. Additionally, value engineering should be viewed as a continuous improvement process, with regular reviews and updates to ensure that cost-saving measures are implemented effectively and that the project remains aligned with its objectives and stakeholders' expectations.

Value engineering is a proactive approach to project management that seeks to maximize value and minimize costs through the identification of cost-saving opportunities. By encouraging collaboration, creativity, and innovation, value engineering empowers project teams to optimize efficiency and resource utilization while delivering high-quality projects that meet or exceed client expectations.

Life-cycle cost analysis

Life-cycle cost analysis is a comprehensive method used in project management to evaluate the total cost of ownership of a project over its entire life span. Unlike traditional cost analysis methods, which focus solely on upfront costs, LCCA takes into account all costs associated with a project, including acquisition, operation, maintenance, and disposal costs. By considering long-term costs and benefits, LCCA enables project stakeholders to make informed decisions that optimize value and minimize total cost of ownership over the project's life cycle.

The process of conducting a life-cycle cost analysis typically involves several key steps:

- **Identifying costs**: The first step in LCCA is to identify all costs associated with the project over its entire life span. This includes not only initial acquisition costs but also ongoing operating and maintenance costs, as well as any costs associated with disposal or decommissioning at the end of the project's life cycle.

- **Estimating costs**: Once costs have been identified, the next step is to estimate the magnitude of each cost component. This may involve gathering data from various sources, such as historical records, industry benchmarks, or expert opinions, to develop accurate cost estimates for each phase of the project's life cycle.

- **Discounting future costs**: Future costs are discounted to account for the time value of money, reflecting the principle that a dollar received or spent in the future is worth less than a dollar received or spent today. Discounting allows for a fair comparison of costs incurred at different points in time and ensures that all costs are expressed in present value terms.

- **Analyzing alternatives**: LCCA involves evaluating different alternatives or options to determine which option provides the lowest total cost of ownership over the project's life cycle. This may include comparing different design alternatives, construction methods, materials, or technologies to identify the most cost-effective solution.

- **Considering non-monetary factors**: In addition to monetary costs, LCCA may also consider non-monetary factors such as environmental impact, social considerations, and risk factors that may affect the overall value and sustainability of the project.

- **Making informed decisions**: Based on the results of the life-cycle cost analysis, project stakeholders can make informed decisions about project design, construction, and management that optimize value and minimize total cost of ownership over the project's life cycle.

By considering long-term costs and benefits, life-cycle cost analysis enables project stakeholders to make informed decisions that optimize value and minimize total cost of ownership over the project's life cycle. By taking a holistic view of costs and benefits, LCCA helps ensure that projects are economically viable, environmentally sustainable, and socially responsible, delivering maximum value to stakeholders and society as a whole.

Building information modeling and its role in estimating

Building Information Modeling is a sophisticated digital technology that revolutionizes the way construction projects are planned, designed, constructed, and managed. BIM serves as a comprehensive digital representation of a building or infrastructure project, encompassing both the physical and functional characteristics of the project throughout its life cycle. One of the key roles of BIM in the construction industry is its significant impact on the estimating process.

- **Virtual modeling**: BIM enables the creation of highly detailed virtual models of the project, including all its components, systems, and materials. These models provide a comprehensive visual representation of the project, allowing estimators to accurately quantify and assess the various elements of the construction, from structural components to mechanical systems.

- **Quantification and takeoffs**: BIM facilitates automated quantification and takeoff processes by extracting relevant data directly from the virtual model. Estimators can generate accurate quantity takeoffs for materials, labor, and equipment based on the information embedded within the BIM model, significantly reducing manual effort and errors associated with traditional takeoff methods.

- **Parametric modeling**: BIM supports parametric modeling, which allows for the creation of intelligent objects with predefined properties and relationships. Estimators can leverage parametric modeling to explore different design alternatives and instantly evaluate their cost implications, enabling informed decision-making during the design development phase.

- **Cost estimation integration**: BIM seamlessly integrates with cost estimation software, enabling estimators to link the virtual model directly to cost databases and estimating tools. This integration facilitates real-time cost analysis and allows for the automatic updating of cost estimates as the design evolves, ensuring accuracy and consistency throughout the estimating process.

- **Visualizations and simulations**: BIM enables the creation of high-quality visualizations and simulations that enhance communication and collaboration among project stakeholders. Estimators can use these visualizations to present cost estimates in a clear and compelling manner, helping clients and other stakeholders better understand the cost implications of design decisions.

- **Clash detection and coordination**: BIM supports clash

detection and coordination by identifying potential conflicts or clashes between different building systems or components. Estimators can use clash detection tools to anticipate and resolve issues during the estimating phase, minimizing costly rework and delays during construction.

BIM plays a pivotal role in the estimating process by providing a digital platform for accurate quantification, cost analysis, and decision support. By leveraging the capabilities of BIM, estimators can streamline their workflows, improve accuracy, and enhance collaboration, ultimately leading to more efficient and cost-effective construction projects.

Examples of BIM application

Design phase

- Architects use BIM to create 3D models of the building design, incorporating architectural elements such as walls, floors, and windows.

- Structural engineers use BIM to design structural systems, including beams, columns, and foundations, and ensure they integrate seamlessly with the architectural design.

- MEP (Mechanical, Electrical, Plumbing) engineers use BIM to design HVAC, electrical, and plumbing systems, ensuring they are coordinated with the architectural and structural elements.

Coordination and clash detection

- BIM models are integrated to identify clashes between different building systems, such as HVAC ducts conflicting with structural elements or electrical conduits interfering with plumbing lines.

- Clash detection tools within BIM software highlight these conflicts, allowing project teams to resolve them before construction begins, thereby reducing costly rework and delays.

Quantity takeoff and cost estimation

- Estimators extract quantities directly from the BIM model, including materials, labor, and equipment, streamlining the quantity takeoff process.
- BIM software integrates with cost estimation tools to generate accurate cost estimates based on the quantities extracted from the model, providing stakeholders with reliable cost projections.

Visualization and communication

- High-quality renderings and visualizations generated from the BIM model are used to communicate design intent to clients, stakeholders, and regulatory authorities.
- 4D BIM adds a time dimension to the model, allowing stakeholders to visualize construction sequences and project phasing, aiding in project planning and scheduling.

Construction planning and sequencing

- Construction managers use BIM to develop detailed construction plans and sequences, visualizing how different components will come together during construction.
- 4D BIM models simulate construction activities over time, helping project teams identify potential conflicts or sequencing issues before they occur on-site.

Facilities management

- BIM models serve as comprehensive digital twins of the built environment, providing facility managers with detailed information about building components, systems, and maintenance requirements.

- Asset data stored within the BIM model can be accessed throughout the building's lifecycle, aiding in maintenance planning, energy management, and space utilization.

These examples illustrate the diverse applications of BIM throughout the lifecycle of a construction project, from design and coordination to construction and facilities management. By leveraging the capabilities of BIM, project teams can enhance collaboration, improve efficiency, and deliver successful outcomes.

CHAPTER 13: STAYING AHEAD OF THE CURVE

Emerging technologies such as Artificial Intelligence and automation are revolutionizing the field of construction estimating, offering new capabilities to streamline processes, improve accuracy, and enhance decision-making. The following are some ways in which AI and automation are transforming construction estimating:

- **AI-powered cost estimation**: AI algorithms analyze historical project data, industry trends, and project specifications to generate accurate cost estimates. Machine learning models can learn from past projects to improve the accuracy of future estimates, taking into account factors such as material costs, labor rates, and regional variations.

- **Automated quantity takeoff**: AI-powered software automates the process of quantity takeoff by analyzing building plans and generating detailed quantity estimates for materials, labor, and equipment. This automation reduces manual effort and errors associated with traditional takeoff methods, saving time and improving accuracy.

- **Advanced data analytics**: AI algorithms analyze vast amounts of project data to identify patterns, trends, and correlations that can inform cost estimates and decision-making. By leveraging advanced data analytics, estimators can gain valuable insights into project cost drivers, risks, and opportunities for optimization.

- **Predictive cost modeling**: AI models use predictive analytics to forecast project costs based on historical data and project characteristics. These models can anticipate potential cost overruns or deviations from the budget, enabling proactive risk management and mitigation strategies.

- **Intelligent bidding systems**: AI-powered bidding platforms use algorithms to analyze bid documents, project specifications, and market conditions to optimize bidding strategies and maximize the chances of winning contracts. These systems automate the bid preparation process, saving time and improving the competitiveness of bids.

- **Automation of routine tasks**: Automation technologies streamline routine tasks such as data entry, document management, and report generation, allowing estimators to focus on value-added activities such as analysis, decision-making, and client interaction. This automation improves productivity and efficiency in the estimating process.

- **Integration with BIM and CAD**: AI and automation technologies integrate seamlessly with Building Information Modeling (BIM) and Computer-Aided Design (CAD) software, enabling seamless data exchange and collaboration between design and estimating teams. This integration improves coordination, accuracy, and efficiency in the estimating process.

AI and automation technologies are transforming construction estimating by enabling advanced cost estimation, automated quantity takeoff, data-driven decision-making, intelligent bidding strategies, and seamless integration with BIM and CAD software. By leveraging these technologies, construction companies can improve accuracy, efficiency, and competitiveness in the estimating process, ultimately leading to better project outcomes and client satisfaction.

Industry trends and innovations impacting estimating practices

Several industry trends and innovations are reshaping estimating practices in the construction industry, driving improvements in accuracy, efficiency, and collaboration. The following are some key trends and innovations impacting estimating practices:

- **Digital transformation**: The construction industry is undergoing a digital transformation, with the widespread adoption of digital tools and technologies such as Building Information Modeling, cloud-based estimating software, and mobile applications. These digital solutions streamline workflows, improve collaboration, and enable real-time access to project data, enhancing the accuracy and efficiency of estimating practices.

- **Data analytics and predictive modeling**: The increasing availability of data and advancements in data analytics are enabling construction companies to leverage predictive modeling techniques to forecast project costs, identify risks, and optimize resource allocation. By analyzing historical data and project characteristics, estimators can make more informed decisions and improve the accuracy of cost estimates.

- **Artificial intelligence and machine learning**: Artificial

Intelligence and Machine Learning algorithms are being used to automate routine tasks, such as quantity takeoff and cost estimation, and to analyze large datasets to identify patterns and trends. AI-powered estimating software can improve the accuracy and speed of cost estimation while reducing manual effort and errors.

- **Integration with Building Information Modeling**: BIM technology is revolutionizing the way construction projects are planned, designed, and executed. Estimating software that integrates with BIM allows estimators to extract quantities directly from the BIM model, improving accuracy and reducing the time required for quantity takeoff. Additionally, BIM enables better coordination between design and estimating teams, leading to more accurate and efficient estimating practices.

- **Prefabrication and modular construction**: Prefabrication and modular construction techniques are gaining popularity due to their potential to reduce construction time, minimize waste, and improve quality control. Estimating practices are adapting to accommodate the unique challenges and opportunities presented by prefabricated and modular construction methods, including estimating costs for off-site fabrication and assembly.

- **Sustainability and green building practices**: There is a growing emphasis on sustainability and green building practices in the construction industry, driven by environmental concerns and regulatory requirements. Estimating practices are evolving to incorporate the costs associated with sustainable design features, energy-efficient materials, and green building certifications, reflecting the increasing demand for sustainable construction solutions.

- **Collaborative workflows and integrated project delivery**: Integrated Project Delivery approaches promote collaboration and integration among project stakeholders, including owners, designers, contractors, and subcontractors. Estimating practices are adapting to support collaborative workflows and shared decision-making processes, enabling more accurate and transparent cost estimation throughout the project lifecycle.

Industry trends and innovations such as digital transformation, data analytics, AI and ML, BIM integration, prefabrication, sustainability, and collaborative workflows are driving significant advancements in estimating practices in the construction industry. By embracing these trends and leveraging innovative technologies, construction companies can improve the accuracy, efficiency, and competitiveness of their estimating processes, ultimately leading to better project outcomes and client satisfaction.

Continuous learning and staying current in the field

Continuous learning and staying current in the field of construction estimating are essential for professional growth, staying competitive, and adapting to industry advancements. Some strategies for continuous learning and staying current in the field include the following:

- **Professional development programs**: Participate in professional development programs, workshops, seminars, and conferences offered by industry organizations, trade associations, and educational institutions. These programs provide opportunities to learn about the latest trends, technologies, and best practices in construction estimating.

- **Continuing education courses**: Enroll in continuing education courses and certifications relevant to construction estimating, such as Certified Professional Estimator (CPE) or Project Management Professional (PMP). These courses help expand knowledge, develop new skills, and demonstrate expertise in the field.

- **Industry publications and journals**: Stay informed about industry trends, case studies, and best practices by reading industry publications, journals, and trade magazines. Subscribing to online platforms, newsletters, and blogs focused on construction estimating can provide valuable insights and updates.

- **Networking and peer learning**: Join professional networking groups, online forums, and social media communities dedicated to construction estimating. Engage with peers, share experiences, and exchange knowledge to learn from others' perspectives and stay abreast of industry developments.

- **Technology training**: Stay current with advancements in estimating software, BIM technology, and digital tools by undergoing regular training and certification programs offered by software vendors and industry organizations. Mastering the latest technology can enhance efficiency, accuracy, and competitiveness in estimating practices.

- **Mentorship and coaching**: Seek mentorship and coaching from experienced professionals in the field of construction estimating. Mentors can provide guidance, advice, and insights based on their industry experience, helping navigate challenges and accelerate professional growth.

- **Participation in industry events**: Attend industry

events, trade shows, and exhibitions related to construction and estimating to network with industry experts, explore new products and technologies, and gain exposure to innovative practices and solutions.
- **Research and self-study**: Dedicate time to independent research and self-study to explore emerging trends, technologies, and methodologies in construction estimating. Utilize online resources, webinars, case studies, and whitepapers to deepen understanding and expand knowledge in specific areas of interest.

By embracing a mindset of continuous learning and staying current with industry developments, construction estimators can enhance their skills, expand their knowledge, and remain competitive in a rapidly evolving field. Continuous learning not only benefits individual professionals but also contributes to the overall advancement and innovation of the construction industry.

Potential impact of AI and automation on job roles within estimating

The integration of Artificial Intelligence and automation technologies into construction estimating processes is poised to have a significant impact on job roles within the estimating profession. While these technologies offer numerous benefits such as increased efficiency, accuracy, and productivity, they also bring about changes in the roles and responsibilities of estimators.

With the automation of routine tasks such as data entry, quantity takeoff, and cost estimation, estimators may find themselves spending less time on manual, repetitive activities. Instead, they can focus on more value-added tasks such as data analysis, interpretation, and decision-making.

AI algorithms can analyze large datasets and extract insights to inform decision-making and improve estimating accuracy. Estimators may need to develop skills in data analysis, interpretation, and visualization to effectively leverage AI-driven insights in their estimating processes.

Estimators will need to adapt to using AI-powered estimating software, automation tools, and digital platforms as part of their daily workflow. This may require training and upskilling to become proficient in using these new tools effectively.

As AI and automation streamline estimating processes, there may be a greater emphasis on collaboration and communication among project stakeholders. Estimators may need to work closely with design teams, contractors, and clients to ensure accurate data inputs, validate assumptions, and align cost estimates with project objectives.

With routine tasks automated, estimators have the opportunity to transition to more strategic roles within their organizations. They can take on responsibilities such as strategic planning, risk management, value engineering, and client relationship management, leveraging their expertise to add value at a higher level.

Estimators will need to stay current with advancements in AI, automation, and construction technology to remain competitive in the field. Continuous learning and professional development will be essential to adapt to evolving job requirements and emerging trends in estimating practices.

As job roles evolve with the adoption of AI and automation, there may be opportunities for reskilling and upskilling to acquire new competencies and stay relevant in the workforce. Estimators may need to pursue additional training or certifications in areas such as data analytics, AI programming, and digital transformation.

Overall, while AI and automation technologies have the

potential to transform the role of estimators, they also present opportunities for career growth, innovation, and professional development. By embracing these technologies and adapting to changing job roles, estimators can remain valuable contributors to the construction industry and drive positive outcomes in estimating practices.

APPENDIX

Glossary

- **4D BIM**: Building Information Modeling integrated with scheduling data to visualize construction sequences and project phasing over time.

- **Assembly cost**: The cost associated with pre-defined assemblies or components used in construction, often based on historical data or industry standards.
- **Bid**: A formal proposal submitted by a contractor or subcontractor to perform work on a construction project, including a breakdown of costs and pricing.

- **Building Information Modeling**: A digital representation of a construction project that integrates 3D models with data to facilitate design, coordination, and visualization.

- **Cloud-based software**: Software applications and services that are hosted and accessed over the internet, offering scalability, accessibility, and collaboration features for construction estimating.

- **Clash detection**: The process of identifying conflicts or clashes between different building systems or components in a BIM model to prevent construction errors and rework.

- **Construction estimating**: The process of calculating the costs associated with a construction project, including materials, labor, equipment, and overhead expenses.

- **Contingency**: An allowance included in a construction estimate to cover unforeseen events or changes in project scope.

- **Desktop software**: Software applications installed and run on a local computer, offering robust features and customization options for construction estimating.

- **Estimate**: An approximation of the total cost of a construction project based on available information, including materials, labor, overhead, and profit margin.

- **Integrated project delivery**: A collaborative approach to project delivery that involves all project stakeholders working together from the early stages to optimize project outcomes and minimize risks.

- **Labor rate**: The cost per hour or unit of time for labor, including wages, benefits, and overhead expenses.

- **Life-cycle cost analysis**: An evaluation of the total cost of ownership of a construction project over its entire life cycle, including acquisition, operation, maintenance, and disposal costs.

- **Material cost**: The cost of materials used in construction, including raw materials, components, and supplies.

- **Overhead costs**: Indirect costs associated with running a construction business, including administrative expenses, utilities, and equipment maintenance.

- **Parametric estimating**: A cost estimation method that uses historical data and project characteristics to generate estimates based on predefined parameters or formulas.

- **Prefabrication**: The process of manufacturing

components or assemblies off-site and transporting them to the construction site for installation, reducing construction time and labor costs.

- **Professional estimator**: A trained and experienced individual responsible for preparing accurate and competitive construction estimates based on project requirements and industry standards.
- **Profit margin**: The percentage of revenue or total cost added to a bid or estimate to account for profit and risk.
- **Quantity surveying**: The profession of estimating and managing construction costs, including quantity takeoffs, cost analysis, and procurement.
- **Risk management**: The process of identifying, assessing, and mitigating risks associated with a construction project to minimize potential losses and disruptions.
- **Sustainability**: The practice of designing, constructing, and operating buildings in an environmentally responsible and resource-efficient manner to minimize environmental impact.
- **Takeoff**: The process of quantifying and measuring materials, labor, and equipment required for a construction project based on drawings and specifications.
- **Unit cost**: The cost per unit of measurement (e.g., per square foot, per cubic yard) for materials, labor, or equipment used in construction.
- **Value engineering**: A systematic approach to optimize the value of a construction project by identifying cost-saving opportunities without sacrificing quality or performance.

Sample estimating template

1. Quantity takeoff templates

- Detailed breakdown of materials, labor, and equipment required for each phase of construction.
- Includes columns for item descriptions, quantities, unit prices, and total costs.
- Can be customized based on project specifications and requirements.

2. Cost estimate worksheets

- Comprehensive spreadsheet for estimating project costs, including materials, labor, overhead, and profit margin.
- Organized by cost category (e.g., materials, labor, equipment) for easy reference.
- Includes formulas for automatic calculation of totals and subtotals.

3. Unit price analysis forms

- Analysis of unit prices for materials, labor, and equipment based on historical data and current market rates.
- Helps in comparing and evaluating unit prices from multiple suppliers or subcontractors.
- Includes fields for unit descriptions, quantities, unit prices, and total costs.

4. Bid proposal templates

- Formal proposal document submitted to clients or project owners outlining the scope of work, schedule, and pricing.
- Includes sections for project overview, scope of work, pricing breakdown, terms and conditions, and signatures.
- Can be customized with company logo, branding, and project-specific details.

5. Labor productivity worksheets

- Analysis of labor productivity rates for different tasks and activities on a construction project.
- Helps in estimating labor hours, crew sizes, and resource allocation based on historical data and industry standards.
- Includes fields for task descriptions, estimated hours, crew sizes, and productivity rates.

6. Material cost analysis forms

- Analysis of material costs based on quantities, unit prices, and waste factors.
- Helps in estimating material costs for different construction activities and phases.
- Includes columns for material descriptions, quantities, unit prices, waste factors, and total costs.

7. Subcontractor quote request forms

- Request for quotes sent to subcontractors for pricing on specific scopes of work or trade packages.
- Includes sections for project details, scope of work, bid instructions, and contact information.

- Can be customized for different trades and subcontractor specialties.

8. Change order forms

- Documentation of changes to the original scope of work, schedule, or pricing during the course of a construction project.
- Includes sections for description of change, rationale, cost impact, and approval signatures.
- Helps in maintaining accurate records of project changes and managing project costs.

These sample estimating forms and templates are designed to provide a foundation for creating customized documents tailored to the specific needs and requirements of each construction project. They can be adapted, modified, and expanded as needed to suit different project types, sizes, and complexities.

Resources for further learning

Recommended reading

- "Construction Estimating: Principles and Practices" by Stephen J. Peterson and Frank R. Dagostino
- "Estimating in Building Construction" by Frank R. Dagostino, Steven J. Peterson, and S. Robert L. Peurifoy
- "RSMeans Building Construction Cost Data" by RSMeans Engineering Staff
- "Construction Project Management" by Frederick Gould and Nancy Joyce
- "Project Management for Engineering and Construction" by Garold D. Oberlender

Online resources

- Construction Dive: A digital media company that covers news and trends in the construction industry, including articles on estimating practices and technologies.
- Construction Executive: A magazine and online platform providing insights and analysis on construction industry trends, including articles on estimating best practices and case studies.
- Construction Estimating Institute: Offers online courses, webinars, and resources for construction estimators looking to improve their skills and stay current with industry trends.
- American Society of Professional Estimators: Provides educational resources, certification programs, and networking opportunities for construction estimators and industry professionals.
- Construction Specifications Institute: Offers educational programs, industry standards, and networking events for professionals involved in construction estimating and project management.

Professional organizations

- American Society of Professional Estimators: A professional organization dedicated to promoting excellence in construction estimating through education, certification, and networking.
- Construction Management Association of America: Provides professional development, certification, and networking opportunities for construction managers, including those involved in estimating.
- National Association of Home Builders: Offers resources, training, and advocacy for professionals in the residential construction industry, including estimators.
- Associated General Contractors of America: Represents

the interests of construction contractors and offers educational programs, safety resources, and industry insights for construction professionals.
- Project Management Institute: Provides certification, training, and resources for project managers, including those involved in construction estimating and project control.

ACKNOWLEDGEMENT

To the readers of this book, striving to master construction estimating, I extend my heartfelt thanks. It is my sincere hope that the knowledge and insights shared within these pages will empower you on this path.

I would like to express my deepest gratitude to everyone who contributed to the creation of this book.

I want to thank my mentors and industry experts whose guidance and wisdom have been invaluable on this journey. In particular, I am grateful to Harlow Parker, whose insights into practical construction estimating were instrumental in shaping the content of this book. Their dedication to excellence in construction estimating continues to inspire me.

I am immensely grateful to my colleagues and peers who have shared their experiences and expertise, enriching the content with diverse perspectives and real-world insights.

A special thank you to the editors who believed in this project and provided invaluable support and feedback throughout the writing process. Your dedication to excellence, especially Carter Moore's attention to detail, has helped bring this book to life.

To my family and friends, thank you for your unwavering support, encouragement, and understanding. Your belief in me has been a constant source of motivation, and I am deeply appreciative of your love and encouragement.

Last but not least, I want to extend my heartfelt thanks to the rest of the contributors who may not be mentioned here by name. Your contributions have played a vital role in making this book a reality.

Thank you all for being a part of this incredible journey.

ABOUT THE AUTHOR

Steven Smith, Ph.d.

Steven Smith is a renowned expert in the field of Construction Management, with a wealth of knowledge and experience spanning both academia and industry. Holding a doctorate in Construction Management, Steven has dedicated his career to advancing the field and contributing to its body of knowledge.

Throughout his academic journey, Steven's passion for understanding the intricacies of construction processes and finding innovative solutions to industry challenges became evident. His doctoral research focused on optimizing project management practices and enhancing productivity in construction projects, leading to a profound understanding of various aspects of construction management and their impact on project success.

www.ingramcontent.com/pod-product-compliance
Lightning Source LLC
Chambersburg PA
CBHW071209240526
45470CB00018B/1643